Lecture Notes in Mathematics

Edited by A. Dold, B. Eckmann and F. Takens

Subseries: Mathematatisches Institut der Universität und
Max-Planck-Institut für Mathematik, Bonn – vol. 14

1400

Uwe Jannsen

Mixed Motives and Algebraic K-Theory

Springer-Verlag

Berlin Heidelberg New York London Paris Tokyo Hong Kong

Author

Uwe Jannsen
Max-Planck-Institut für Mathematik
Gottfried-Claren-Str. 26
5300 Bonn 3, Federal Republic of Germany

Mathematics Subject Classification (1980): Primary: 14A20, 14C30, 14G13, 18F25
Secondary: 12A67, 14C35, 14F15

ISBN 3-540-52260-3 Springer-Verlag Berlin Heidelberg New York
ISBN 0-387-52260-3 Springer-Verlag New York Berlin Heidelberg

Printing and binding: Druckhaus Beltz, Hemsbach/Bergstr.
2146/3140-543210 – Printed on acid-free paper

Preface

This is an almost unchanged version of my 1988 Habilitationsschrift at Regensburg. My original plan was to completely rewrite it for publication; in particular I wanted to make it more readable for the non–expert. Finally I chose to rather publish it like it is than turn it into a long range project. So I have only made some minor corrections and added three appendices. The first one reproduces a letter from S. Bloch to me and the second one consists of an example by C. Schoen. I thank both for the permission to publish this material, and the latter for the effort of rewriting the example, which also figured in a letter to me. The third appendix contains some remarks and complements written in 1989.

Uwe Jannsen
Bonn, November 1989

Introduction

This text consists of three parts. In part I we define a category of mixed motives in the setting of absolute Hodge cycles. In part II we investigate, as general as possible, relations between algebraic cycles, algebraic K-theory, and mixed structures in the cohomology of arbitrary varieties. In part III we present some conjectures on Chern characters from K-theory into ℓ-adic cohomology for varieties over finite fields or global fields, and prove these in some (very) specific cases.

Background The concept of motives [Ma] ,[Kl] , [SR] was introduced by Grothendieck to explain phenomena in different cohomology theories of algebraic varieties in a coherent way, in particular those related to algebraic cycles and weights. For example in both the ℓ-adic and the Hodge theory the cohomology $H^i(X)$ of a smooth projective variety is pure of weight i, the class of an algebraic cycle of codimension j can be interpreted as a morphism from the trivial structure into $H^{2j}(X)(j)$, and the parallel formulation of the conjectures of Hodge and of Tate is that the functor sending a motive to its cohomological realization is fully faithful.

All this only concerns cycles modulo homological equivalence and does not cover singular or non-compact varieties, which often arise in algebraic geometry. Concerning these, Deligne shows in [D5] §10 that cycles homologous to zero give rise to non-trivial extensions of pure structures of different weights - this is called a mixed structure - and in his treatments of Hodge theory and ℓ-adic cohomology [D5] , [D9] shows that the cohomology of arbitrary varieties gives rise to mixed structures, too. Indeed, both facts are directly related, and one expects a description of the whole Chow group and a satisfactory treatment of arbitrary varieties in the setting of a category of mixed motives [Bei 4], [D10]. Finally, work of Beilinson suggests that mixed motives are related to higher algebraic K-theory, like cycles are related to K_o [Bei 1], [Bei 2].

Grothendieck's definition of motives is quite simple, but only gives a satisfactory theory together with the so-called standard conjectures. Deligne has given a "working definition" of motives for absolute Hodge cycles (the latter ones replacing the algebraic cycles in Grothendieck's definition), which often suffices for the applications [DMOS]. An algebraic definition of mixed motives is problematic, since Grothendieck's methods (algebraic correspondences and idempotents) neither apply nor extend in an obvious way.

Part I In §1 we start with the simple but crucial observation that - in the language introduced later - a subrealization of the realization of a motive for absolute Hodge cycles (AH-motive) is a direct factor and hence a submotive. As a corollary we show that there are natural AH-motives associated to modular forms, having as ℓ-adic realizations the representations constructed by Deligne [D1] (Recently, Scholl [Sch 1] constructed these motives algebraically). Another application is the construction of direct factors in the ℓ-adic cohomology.

In §2 we make a precise definition of a category $\underline{R}_k$ in which the realizations of AH-motives over a field k live, by defining a bigger category $\underline{MR}_k$ of mixed realizations, in which also mixed structures are allowed. These obviously are Tannakian categories, and we study some of their formal properties.

In §3 we prove that for a smooth variety U over a field k of characteristic zero its ℓ-adic, deRham and Betti cohomologies define an object H(U) in $\underline{\underline{MR}}_k$. The techniques applied here are all taken from papers of Deligne, the main point consisting in showing that one has a weight filtration in each theory which is compatible with the comparison isomorphisms, and that the pure quotients are AH-motives.

In §4 the category $\underline{\underline{MM}}_k$ of mixed motives over k is defined as the Tannakian subcategory of $\underline{\underline{MR}}_k$ generated by the H(U) . We prove that Deligne's category $\underline{\underline{M}}_k$ can be identified with the Tannakian subcategory generated by the realizations of smooth, projective varieties, and can be identified with the full subcategory of pure objects in $\underline{\underline{MM}}_k$. This gives a simpler definition of $\underline{\underline{M}}_k$ than the original one, avoiding the processes of taking the pseudo-abelian hull, inverting the Lefschetz object and changing the commutation constraints. If G and MG are the associated "Galois groups" of the neutral Tannakian categories $\underline{\underline{M}}_k$ and $\underline{\underline{MM}}_k$ (for some fibre functor given by Betti cohomology), then the embedding $\underline{\underline{M}}_k \hookrightarrow \underline{\underline{MM}}_k$ defines a homomorphism MG → G , and the above is reflected in an exact sequence of pro-algebraic groups

$$1 \to U \to MG \to G \to 1 ,$$

with connected, pro-unipotent U, identifying G with the maximal pro-reductive quotient of MG .

Part II §5 is, except for theorems 5.13 and 5.15 (comparing $\mathcal{O}(X)^{\times}$ with Deligne cohomology $H^1_{\mathcal{D}}(X,\mathbb{Z}(1))$ or étale cohomology $H^1_{et}(X,\mathbb{Z}_\ell(1))$), mainly motivational. The conjectures stated here for the smooth case are contained in those formulated later for arbitrary varieties.

In §6 a very important tool appears, the notion, due to Bloch and Ogus [BO],of a twisted Poincaré duality theory, axiomatizing the aspects of a cohomology theory and an associated homology theory.

In this setting the "Poincaré duality" is an isomorphism

$$(0.1) \qquad H^i(X,j) \xrightarrow{\sim} H_{2d-i}(X,d-j) \quad , \quad d = \dim X \ ,$$

between cohomology and homology for smooth X . We define a version with values in a tensor category, also introducing the concept of weights modeled after the situation for mixed Hodge structures or mixed ℓ-adic sheaves. After discussing ℓ-adic, deRham and Betti-cohomology we prove - extending the results in part I - that there is a Poincaré duality theory with values in $\underline{MR}_k$.

In §7 we propose how to extend the conjectures of Hodge and Tate to arbitrary varieties. The basic observation is that the right setting is the homology, the classical formulations being reobtained by (0.1). We show that this Hodge conjecture is true if and only if the classical Hodge conjecture is, and that the same is basically true for the Tate conjectures.

In §8 we recall some properties of Chern characters and Riemann-Roch transformations assuring that the maps

$$(0.2) \qquad H_a^{\mathcal{M}}(X,\mathbb{Q}(b))\otimes\mathbb{Q}_\ell \to H_a^{\text{ét}}(X\times_k\bar{k},\mathbb{Q}_\ell(b))^{G_k} \quad , \quad \text{char } k \neq \ell \ ,$$

$$(0.3) \qquad H_a^{\mathcal{M}}(X,\mathbb{Q}(b)) \to \Gamma_{\mathcal{H}} H_a(X(\mathbb{C}),\mathbb{Q}(b)) \quad , \quad k = \mathbb{C} \ ,$$

(where $H^{\mathcal{M}}$ is the motivic homology defined by Beilinson via $K'_*(X)$ and $\Gamma_{\mathcal{H}}$ denotes the group of Hodge cycles), satisfy all functorialities of morphisms of Poincaré duality theories. We state conjectures on the surjectivity of (0.2) and (0.3) and extend theorems 5.13 and 5.15 to arbitrary varieties, thus proving the conjectures for curves.

In §9 we discuss relations between extensions of realizations and algebraic cycles homologous to zero. As a consequence we show why a naive extension of the conjectures of Hodge and Tate to the surjectivity of (0.2) and (0.3) for arbitrary $a,b \in \mathbb{Z}$ is false. In particular, this disproves a Hodge-theoretic conjecture by Beilinson [Bei 2] . We deduce the counterexample from examples of Mumford on the non-injectivity of the Abel-Jacobi map

$$CH^j(X)_0 \to H^{2j-1}(X,\mathbb{C})/H^{2j-1}(X,\mathbb{Z}(j)) + F^j \ .$$

Then we extend everything to the ℓ-adic Abel-Jacobi maps

$$CH^j(X)_o \to H^1_{cont}(G_k, H^{2j-1}(X\times_k\bar{k}, \mathbb{Z}_\ell(j))) ,$$

by using results of Bloch [Bl 1] .

In §10 we extend Bloch's results to higher-dimensional varieties and show that Abel-Jacobi maps are non-injective quite principally, for any reasonable Poincaré duality theory - provided the base field contains too many parameters. The main theme of our conjectures, and of several conjectures of Bloch and Beilinson, is that the situation is different for finite fields, global function fields, and number fields.

In §11 we recall some ideas of Beilinson on mixed motives [Bei 4]. We stress the fact that his philosophy of mixed motivic sheaves would imply some quite explicit conjectures - extending earlier ones by Bloch - on the structure of Chow groups of smooth projective varieties over arbitrary fields. I think these should be regarded as an extension of Grothendieck's standard conjectures to the whole Chow group. We remark that they would follow from the injectivity of some cycle map.

Part III Our basic conjecture for varieties over finite fields is that here (0.2) is an isomorphism. In §12 we prove it in some cases and show that it would follow from several "classical" conjectures on smooth, projective varieties, at least if we assume a weak form of resolution of singularities. The conjecture would imply a description of motivic homology of arbitrary varieties X over arbitrary fields of positive characteristic, by writing $X = \varprojlim_\alpha X_\alpha$, with varieties X_α over $\mathbb{F}_p$ and flat transition maps, since $H^{\mathcal{M}}_a(X,\mathbb{Q}(b)) = \varinjlim_\alpha H^{\mathcal{M}}_{a'}(X_\alpha,\mathbb{Q}(b'))$. We explain this in more detail for the case of a global function field k . Note that we need non-proper X_α even for a smooth, projective X , and observe the similarities and the differences to the approach of Artin and Tate in [D.E.] .

We don't have a similarly general conjecture for number fields, but in §13 we discuss a conjecture on the bijectivity of

(0.4) $H_a^M(X,\mathbb{Q}(b))\otimes\mathbb{Q}_\ell \to \tilde{H}_a^{et}(X,\mathbb{Q}_\ell(b))$,

(where $\tilde{H}_*^{et}$ is a certain modified étale homology) in the "stable range" $a > \dim X + b$. This is related to certain Galois cohomological investigations in [J3] .

The extreme counterpart of pure structures are mixed structures whose pure pieces are as simple as possible, i.e., Tate objects, so that only mixed phenomena remain. In §14 we define a class of varieties (containing those stratified by linear spaces, like Grassmannians or flag varieties) with this property, and prove most of our conjectures for these varieties.

Final remarks and acknowledgements

I learnt about motives for absolute Hodge cycles in inspiring lectures by G. Anderson (Harvard 1983/84), and my own investigations were started by a question of N. Schappacher whether the realizations for modular forms come from such motives (see §1). A. Scholl brought my attention to the paper by Bloch and Ogus, and communicated to me some ideas on K-homology and extension classes (cf. §6). It is a pleasure to thank them for this inspiration and the latter two for further discussions.

The first four chapters exist in this form since end of 1985 and were communicated to a few mathematicians. It should be noted that a construction similar to our category $\underline{MR}_k$ also appears in a recent paper by Deligne. It will be clear to the reader how much parts II and III are influenced by work and ideas of Bloch and Beilinson, but I would also like to stress the influence of Deligne's work on 1-motives [D5] and his reinterpretation of Beilinson's ideas in [D10] .

I would like to thank J. Neukirch heartily for his constant

enthusiasm and encouragement, and all friends in Regensburg for their interest and support. Also I thank the Max-Planck-Institut at Bonn, where this program was started and where the final part was written. Special thanks go to K. Deutler, M. Grau and H. Wolf-Gazo from the MPI, and in particular to M. Hertl from Regensburg for a phantastic typing under big pressure of time.

Contents

PART I

MIXED MOTIVES FOR ABSOLUTE HODGE CYCLES

§1. Some remarks on absolute Hodge cycles

Let k be a field of characteristic zero, which is embeddable in $\mathbb{C}$. Fix an algebraic closure $\bar{k}$ of k and let $G_k = \mathrm{Gal}(\bar{k}/k)$. In the following we deal with motives for absolute Hodge cycles as defined by Deligne in [D6], see also [DMOS]II §6, in particular we use similar notations as in these references. Then a motive M over k has realizations

$H_{DR}(M)$ - a k-vector space with a descending filtration F^p

$H_l(M)$ - (for each prime number l) a $\mathbb{Q}_l$-vector space, on which G_k acts continuously,

$H_\sigma(M)$ - (for each embedding $\sigma: k \hookrightarrow \mathbb{C}$) a $\mathbb{Q}$-vector space with a Hodge structure on $H_\sigma(M) \otimes \mathbb{R}$, i.e., a $\mathbb{Q}$-Hodge structure,

all of the same finite dimension. Furthermore, there are comparison isomorphisms

$$I_{\infty,\sigma} : H_\sigma(M) \otimes_{\mathbb{Q}} \mathbb{C} \xrightarrow{\sim} H_{DR}(M) \otimes_{k,\sigma} \mathbb{C}$$

and

$$I_{l,\bar{\sigma}} : H_\sigma(M) \otimes_{\mathbb{Q}} \mathbb{Q}_l \xrightarrow{\sim} H_l(M)$$

for each extension $\bar{\sigma} : \bar{k} \hookrightarrow \mathbb{C}$ of σ.

If X is a smooth projective variety over k and $n \geq 0$ an integer, the motive $M = h^n(X)$ is given by the realizations

$$H_{DR}(M) = H^n_{DR}(X) = H^n_{DR}(X/k) \qquad \text{(de Rham cohomology)}$$

$$H_l(M) = H^n_l(X) = H^n_{et}(X\times_k\bar{k},\mathbb{Q}_l) \qquad \text{(l-adic cohomology)}$$

$$H_\sigma(M) = H^n_\sigma(X) = H^n(X\times_{k,\sigma}\mathbb{C},\mathbb{Q}) \qquad \text{(singular cohomology)} .$$

The comparison isomorphisms are obtained from the canonical ones between the cohomology theories of the variety $\sigma X = X \times_{k,\sigma} \mathbb{C}$ over $\mathbb{C}$. Namely $I_{1,\bar{\sigma}}$ is given by

$$H^n(X\times_{k,\sigma}\mathbb{C},\mathbb{Q}_1) \xrightarrow[\sim]{\text{can}} H^n_{et}(X\times_{k,\sigma}\mathbb{C},\mathbb{Q}_1) \xrightarrow[\sim]{\bar{\sigma}^*} H^n_{et}(X\times_k\bar{k},\mathbb{Q}_1)$$

and $I_{\infty,\sigma}$ is induced by

$$H^n(\sigma X,\mathbb{C}) \xrightarrow[\sim]{\text{can}} H^n_{DR}(\sigma X/\mathbb{C}) .$$

If we let $h(X) = \bigoplus_{n=o}^{\dim X} h^n(X)$, any motive M is a direct summand of $h(X)(m)$, the m-fold Tate-twist of $h(X)$, for some smooth projective X and some $m \in \mathbb{Z}$.

The following lemma, which describes the possible summands, is rather easy but very important for the following.

<u>1.1. Lemma</u> Let M be a motive over k. Suppose given a k-subspace $U_{DR} \subseteq H_{DR}(M)$, for each 1 a $\mathbb{Q}_1$-subspace $U_1 \subseteq H_1(M)$, which is a G_k-submodule, and for each $\sigma: k \hookrightarrow \mathbb{C}$ a $\mathbb{Q}$-subspace $U_\sigma \subseteq H_\sigma(M)$, which is a sub-$\mathbb{Q}$-Hodge structure, such that these subspaces correspond under the comparison isomorphisms. Then there is a decomposition $M = M_1 \oplus M_2$ in motives such that $U_\alpha = H_\alpha(M_1) \subseteq H_\alpha(M)$ where α runs through the indices DR, 1 and σ.

<u>Proof</u> As the subspaces U_α are compatible with the weight gradings (this is implicit in the statement that the U_σ are sub-$\mathbb{Q}$-Hodge structures), we may assume M pure of weight r, say. Then there exists a morphism of motives

$$\Psi : M \to \check{M}(-r) \qquad (\check{M} = \text{dual of } M)$$

giving rise to non-degenerate pairings for $\alpha \in \{DR,1,\sigma\}$

$$\psi_\alpha : H_\alpha(M) \otimes H_\alpha(M) \to H_\alpha(1(-r)) = \begin{cases} k & \alpha = DR \\ \mathbb{Q}_1(-r) & \alpha = 1 \\ \mathbb{Q}(-r) & \alpha = \sigma \end{cases}$$

which are compatible with the various structures like G_k-action for $\alpha = 1$ and Hodge structure for $\alpha = \sigma$ etc., and correspond under the comparison isomorphisms. Moreover, the ψ_σ induce polarizations of real Hodge structures.

$$H_\sigma(M)\otimes\mathbb{R} \otimes H_\sigma(M)\otimes\mathbb{R} \rightarrow \mathbb{R}(-r)$$

In fact, to fix ideas we may assume - by twisting with powers of the Tate motive and adding other motives - that M is $h^r(X)$ for a smooth projective variety X of dimension d over k. Then by using a very ample divisor and the hard Lefschetz theorem one constructs an absolute Hodge cycle in $C_{AH}^{2d-r}(X\times X)$ giving a homomorphism

$$\Phi : h^r(X) \rightarrow h^{2d-r}(X)(d-r) \quad ,$$

the motivic version of the "*-operator" in Hodge theory, see [DMOS] II 6.2. The pairings ψ_α above are then obtained by combining with the Poincaré pairings

$$H_\alpha^r(X) \otimes H_\alpha^{2d-r}(X) \rightarrow H_\alpha^{2d}(X) \overset{tr}{\underset{\sim}{\rightarrow}} H_\alpha(1(-d))$$

and twists by $d-r$. Or: the Poincaré pairings give an isomorphism $h^{2d-r}(X)(d-r) \underset{\sim}{\rightarrow} h^r(X)^\vee(-r)$, whose composition with Φ is Ψ.

Let V_{DR}, V_1 and V_σ be the orthogonal complements of U_{DR}, U_1 and U_σ, respectively, with respect to the pairings ψ_{DR}, ψ_1 and ψ_σ. By the compatibility of the ψ_α these spaces then correspond under the comparison isomorphisms. Also the V_α are substructures of the $H_\alpha(M)$ like the U_α : the G_k-invariance of V_1 follows from the G_k-invariance of U_1 and ψ_1, and V_σ is a sub-$\mathbb{Q}$-Hodge structure, as ψ_σ is a polarization of $\mathbb{Q}$-Hodge structures. This also shows that $U_\sigma \cap V_\sigma = 0$ (compare Deligne's argument [D4] p. 44, that any sub-structure of a polarized $\mathbb{Q}$-Hodge structure is a direct factor): one has $(2\pi i)^r \psi_\sigma(x,Cx) > 0$ for all $0 \neq x \in H_\sigma(M) \otimes \mathbb{R}$, where C is the Weil operator: $C = i \in S(\mathbb{R}) = \mathbb{C}^{\times}$ acting on every $\mathbb{R}$-Hodge structure, see [D4] (2.1.14). As C respects the sub-Hodge structure $U_\sigma \otimes \mathbb{R}$ we conclude $U_\sigma \otimes \mathbb{R} \cap (U_\sigma \otimes \mathbb{R})^{\perp} = 0$ as claimed. By the comparison isomorphisms we also get $U_l \cap V_l = 0$ and $U_{DR} \cap V_{DR} = 0$. The decompositions $H_\alpha(M) = U_\alpha \oplus V_\alpha$ then induce endomorphisms

$$p_\alpha : H_\alpha(M) \xrightarrow{\text{projection}} U_\alpha \to H_\alpha(M)$$

for $\alpha \in \{DR, l, \sigma\}$, which are compatible with the various structures and the comparison isomorphisms, as this is the case for the U- and V-spaces. Therefore the family of the p_α gives an element $p \in \mathrm{End}(M)$ (see [DMOS]II 6.7 (g) or 6.1 for $M = h(X)$, note that p_{DR} respects the Hodge filtration as it is compatible with p_σ and p_σ is a homomorphism of Hodge structures), which is a projector and gives the wanted decomposition by taking $M_1 = \mathrm{Im}\, p$ and $M_2 = \mathrm{Im}(1-p)$; for $M = h(X)$ we have $M_1 = (h(X), p)$ in the notation of [DMOS].

<u>1.2. Corollary</u> If X, Y are smooth varieties over k with X projective, then for any morphism $f: Y \to X$ and $g: X \to Y$ the kernel of

$$f^*_\alpha : H^r_\alpha(X) \to H^r_\alpha(Y) \qquad \alpha \in \{DR, l, \sigma\}$$

is represented by a motive $\mathrm{Ker}\, f^* \subseteq h^r(X)$ and the image of

$$g^*_\alpha : H^r_\alpha(Y) \to H^r_\alpha(X) \qquad \alpha \in \{DR, l, \sigma\}$$

is represented by a motive $\mathrm{Im}\, g^* \subseteq h^r(X)$, and these are direct factors of $h^r(X)$.

<u>Proof</u> The cohomology groups $H^r_\sigma(Y)$ have mixed $\mathbb{Q}$-Hodge structures, and f^*_σ and g^*_σ are morphisms of mixed $\mathbb{Q}$-Hodge structures [D4]. So $\mathrm{Ker}\, f^*_\sigma$ and $\mathrm{Im}\, g^*_\sigma$ are (pure) sub-$\mathbb{Q}$-Hodge structures

of the pure, polarized $\mathbb{Q}$-Hodge structures $H^r_\sigma(X)$. Ker f^*_α and Im g^*_α in the other realizations correspond to Ker f^*_σ and Im g^*_σ under the comparison isomorphisms, as these are functorial and also exist for Y , and of course in the l-adic realizations one gets G_k-invariant subspaces. So we can apply the lemma (with U_α = Ker f^*_α or Im g^*_α) .

In particular we get a result which should be true more generally by a conjecture of Grothendieck-Serre on the semi-simplicity of the action of G_k on the l-adic cohomology.

<u>1.3. Corollary</u> In the situation above, the kernel of

$$f^*_l : H^r_l(X) \to H^r_l(Y)$$

and the image of

$$g^*_l : H^r_l(Y) \to H^r_l(X)$$

are direct factors of $H^r_l(X)$ as G_k-modules .

Of course, similar considerations apply to other natural maps like Gysin maps or the canonical map

$$H^r_c(U) \to H^r(X)$$

of the cohomology with compact support of an open subvariety U of X into the cohomology of a smooth projective variety X . This is needed in the proof of the next corollary.

<u>1.4. Corollary</u> The realizations attached to an elliptic modular form f by Deligne ([D6] §7) belong to a motive $M(f)$.

<u>Proof</u> Let f be a new form of weight $k+2$ $(k \geq 0)$, conductor N and character ε for

$$\Gamma_1(N) = \{ \begin{pmatrix} a & b \\ c & d \end{pmatrix} \in SL_2(\mathbb{Z}) \mid \begin{pmatrix} a & b \\ c & d \end{pmatrix} \equiv \begin{pmatrix} 1 & * \\ 0 & 1 \end{pmatrix} \bmod N \} .$$

There is a smooth projective curve $X_1(N)$ over $\mathbb{Q}$ and an open subvariety

$$j: Y_1(N) \hookrightarrow X_1(N)$$

such that the $\mathbb{C}$-valued points can be identified with

$$\Gamma_1(N)\backslash\mathfrak{h} \hookrightarrow \overline{\Gamma_1(N)\backslash\mathfrak{h}} = \text{compactification by adding the cusps} ,$$

where $\mathfrak{h}$ is the Poincaré upper halfplane.

Let $N \geq 3$; then there is the universal elliptic curve

$$g: E \to Y_1(N) ,$$

and Deligne describes the realizations of $M(f)$ as parts of the "universal cohomology"

$$H^1(X_1(N) , j_* \mathrm{Sym}^k(R^1 g_* \mathbb{Q}))$$

(i.e., one has to form the l-adic, de Rham and singular versions of this cohomology), namely as kernel of $T_n - a_n$ for all n prime to N , where the T_n are the Hecke correspondences acting on the cohomology and $f(z) = \sum_{n \geq 1} a_n q^n$, $q = e^{2\pi i z}$. If the a_n are not in $\mathbb{Q}$, one has to take the kernel in the following sense: Let T be the $\mathbb{Q}$-algebra generated by the T_n and $E = \mathbb{Q}(a_1, a_2, \ldots)$, then we have a morphism $T \twoheadrightarrow E$ by $T_n \mapsto a_n$. If $\mathfrak{a}$ is the kernel of this morphism, define the realizations of $M(f)$ as the part annihilated by $\mathfrak{a}$.

By the commutative diagram

$$\begin{array}{ccc} H^1_c(Y_1(N), \mathrm{Sym}^k(R^1 g_* \mathbb{Q})) & \twoheadrightarrow & H^1(X_1(N), j_* \mathrm{Sym}^k(R^1 g_* \mathbb{Q})) \\ & \searrow^{\varphi} & \downarrow \\ & & H^1(Y_1(N), \mathrm{Sym}^k(R^1 g_* \mathbb{Q})) \end{array} ,$$

in which H^1_c denotes cohomology with compact support and the maps are the canonical ones, one can also define the realizations of $M(f)$ to be the kernel of the $T_n - a_n$ in the parabolic cohomology

$$H^1_p(Y_1(N),\mathrm{Sym}^k(R^1g_*\mathbb{Q})) = \mathrm{Im}(H^1_c(Y_1(N),\ldots) \overset{\varphi}{\to} H^1(Y_1(N),\ldots)).$$

$\mathrm{Sym}^k(R^1g_*\mathbb{Q})$ is a direct factor of $(R^1g_*\mathbb{Q})^{\otimes k}$ which in turn is a direct factor of $R^k(g_k)_*\mathbb{Q}$, for

$$g_k : E_k = E \times_{Y_1(N)} \cdots \times_{Y_1(N)} E \to Y_1(N)$$

the k-fold fibre product of g (relative version of the Künneth formula), where by definition $E_o = Y_1(N)$.

Finally the spectral sequence

$$H^p(Y_1(N),R^q(g_k)_*\mathbb{Q}) \Rightarrow H^{p+q}(E_k,\mathbb{Q})$$

degenerates and moreover, as remarked by Lieberman, identifies $H^p(Y_1(N),R^q(g_k)_*\mathbb{Q})$ with the subspace of $H^{p+q}(E_k,\mathbb{Q})$, on which $m \cdot \mathrm{id}_{E_k}$ induces the multiplication by m^q, compare [D1] p. 168. The same is true for the cohomology with compact support.

Altogether the realizations of $M(f)$ are direct factors of the cohomology

$$H^{k+1}_p(E_k,\mathbb{Q}) = \mathrm{Im}(H^{k+1}_c(E_k,\mathbb{Q}) \to H^{k+1}(E_k,\mathbb{Q}))$$

which are defined as the kernel of several algebraic correspondences: the T_n are also defined as correspondences of E and so of E_k, see [D1] (3.16), the subquotient of $H^{k+1}_p(E_k,\mathbb{Q})$ which corresponds to

$$H^1_p(Y_1(N),(R^1g_*\mathbb{Q})^{\otimes k}) \subseteq H^1_p(Y_1(N),R^k(g_k)_*\mathbb{Q})$$

via the spectral sequence can be identified with the subspace of $H^{k+1}_p(E_k,\mathbb{Q})$ where the morphism $m_1\mathrm{id}_E \times \ldots \times m_K\mathrm{id}_E$ $(m_i \in \mathbb{Z})$ induces the multiplication by $m_1 \ldots m_k$, and the part corresponding to $\mathrm{Sym}^k(R^1g_*\mathbb{Q})$ in $(R^1g_*\mathbb{Q})^{\otimes k}$ can be identified by the action of the symmetric group S_K on E_K.

If one likes - and in particular if one does not like to

elaborate the de Rham versions of the above steps - one can take this as the definition: the realizations of $M(f)$ are obtained in $H^{k+1}_{p,\alpha}(E_k) = \mathrm{Im}(H^{k+1}_{\alpha,c}(E_k) \overset{\varphi_\alpha}{\to} H^{k+1}_{\alpha}(E_k))$, for $\alpha \in \{DR, l, \sigma\}$, as the kernel of the $T_n - a_n$, $(m_1 id_E \times \ldots \times m_K id_E)^* - m_1 \ldots m_K$ for sufficiently many $m_i \in \mathbb{Z}$, and $\sigma^* - 1$ for all $\sigma \in S_k \subset \mathrm{Aut}(E_k)$. They are substructures, i.e., G_k-submodules of $H^{k+1}_l(E_k)$, sub-$\mathbb{Q}$-Hodge structures of $H^{k+1}_{\sigma}(E_k)$ etc., and correspond under the comparison isomorphisms, as these also exist for the cohomology with compact support $H^{i}_{\alpha,c}(E_K)$ and are compatible with the φ_α . A definition of algebraic de Rham cohomology with compact support and the comparison isomorphism to singular cohomology can be found in [HL] .

To get a motive we still have to replace E_k by a smooth projective variety. Now there exists a smooth compactification $\bar{E}_k$ of E_k , i.e., a smooth projective variety $\bar{E}_k$ containing E_k as an open subvariety (either by Hironaka's resolution of singularities or by Deligne's direct construction [D1] 5.5, which also works in positive characteristic), and by the commutative diagram

$$\begin{array}{ccc} H^{k+1}_{\alpha,c}(E_k) & \xrightarrow{\psi} & H^{k+1}_{\alpha}(\bar{E}_k) \\ & \searrow & \downarrow \rho \\ & & H^{k+1}_{\alpha}(E_k) \end{array}$$

$H^{k+1}_{\alpha,p}(E_k)$ appears as a subquotient of $H^{k+1}_{\alpha}(\bar{E}_k)$. We remark that by the commutative diagram from Poincaré duality

$$\begin{array}{ccccc} H^{k+1}(\bar{E}_k) & \times & H^{k+1}(\bar{E}_k) & \to & H^{2(k+1)}(\bar{E}_k) \\ \psi \uparrow & & \downarrow \rho & & \uparrow \int \\ H^{k+1}_c(E_k) & \times & H^{k+1}(E_k) & \to & H^{2(k+1)}_c(E_k) \end{array}$$

we have $\mathrm{Im}\ \psi = (\mathrm{Ker}\ \rho)^{\perp}$ (orthogonal complement). This shows

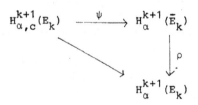

that we could express the subquotient entirely in terms of ρ :
$H_{\rho}^{k+1}(E_k) = \rho(\mathrm{Im}\ \psi) \cong \mathrm{Im}\ \psi/\mathrm{Im}\ \psi \cap \mathrm{Ker}\ \rho = (\mathrm{Ker}\ \rho)^{\perp}/(\mathrm{Ker}\ \rho)^{\perp} \cap \mathrm{Ker}\ \rho$.

These subquotients for all $\alpha \in \{DR, l, \sigma\}$ define a motive by lemma 1.1 (applied twice), and the realizations of $M(f)$ give compatible substructures in all its realizations and so again by lemma 1.1 define a motive, which we now can call $M(f)$.

More explicitely: let $H(M(f))$ be the realizations of $M(f)$ in $H^{k+1}(E_k)$, then $\rho^{-1}(H(M(f)))$ gives a motive by lemma 1.1, $(\mathrm{Ker}\ \rho)^{\perp} \cap \mathrm{Ker}\ \rho$ is a motive, and so $\rho^{-1}(H(M(f)))/(\mathrm{Ker}\ \rho)^{\perp} \cap \mathrm{Ker}\ \rho$ is a motive, which we define to be $M(f)$.

For $N = 1,2$ one gets the motive $M(f)$ from a motive with bigger N' via taking the fixed part under a finite subgroup of $SL_2(\mathbb{Z}/N'\mathbb{Z})$ like in [D1] p. 158. This again gives a motive, compare [DMOS] p. 206.

§2. The category of mixed realizations

The considerations of the previous section suggest to define a category that contains the realizations of motives with all their extra structures and that also covers the cohomology of (smooth) non-proper varieties, which in general gives rise to mixed structures. We do this by formalizing the properties of the realizations and replacing the weight graduation of motives by a weight filtration.

Let again k be a field, which is embeddable in $\mathbb{C}$, $\bar{k}$ an algebraic closure of k, and $G_k = \mathrm{Gal}(\bar{k}/k)$.

<u>2.1. Definition</u> The category $\underline{MR}_k$ of mixed realizations (for absolute Hodge cycles) over k consists of families

$$H = (H_{DR}, H_l, H_\sigma;\ I_{\infty,\sigma}, I_{l,\bar\sigma}) \quad l \text{ prime number}$$
$$\sigma : k \hookrightarrow \mathbb{C}$$
$$\bar\sigma : \bar k \hookrightarrow \mathbb{C}$$

where

a) H_{DR} is a finite-dimensional k-vector space with a decreasing filtration $(F^n)_{n\in\mathbb{Z}}$ (the <u>Hodge filtration</u>) and an increasing filtration $(W_m)_{m\in\mathbb{Z}}$ (the <u>weight filtration</u>).

b) H_l is a finite-dimensional $\mathbb{Q}_l$-vector space with a continuous G_k-action and an increasing filtration $(W_m)_{m\in\mathbb{Z}}$ (the <u>weight filtration</u>), which is G_k-equivariant.

c) H_σ is a mixed $\mathbb{Q}$-Hodge structure, i.e., there is an increasing filtration $(W_m)_{m\in\mathbb{Z}}$ (the <u>weight filtration</u>) on H_σ and a decreasing filtration $(F^n)_{n\in\mathbb{Z}}$ (the <u>Hodge filtration</u>) on $H_\sigma \otimes_{\mathbb{Q}} \mathbb{C}$, which induces a $\mathbb{Q}$-Hodge structure of weight m on $Gr_m^W H_\sigma = W_m H_\sigma / W_{m-1} H_\sigma$, that is $Gr_m^W H_\sigma \otimes \mathbb{C} = \bigoplus_{p+q=m} H_\sigma^{p,q}$ with $\overline{H_\sigma^{p,q}} = H_\sigma^{q,p}$ and $F^p Gr_m^W H_\sigma \otimes \mathbb{C} = \bigoplus_{p' \geq p} H_\sigma^{p',q'}$.

d) $I_{\infty,\sigma} : H_\sigma \otimes_{\mathbb{Q}} \mathbb{C} \xrightarrow{\sim} H_{DR} \otimes_{k,\sigma} \mathbb{C}$ is an isomorphism identifying the filtrations induced by the Hodge filtrations (respectively, the weight filtrations) on both sides.

e) $I_{l,\bar\sigma} : H_\sigma \otimes_{\mathbb{Q}} \mathbb{Q}_l \xrightarrow{\sim} H_l$, for $\sigma = \bar\sigma|_k$, is an isomorphism transforming the weight filtration of H_σ into the weight filtration of H_l, such that for $\rho \in G_k$

$$\begin{array}{ccc} & & H_l \\ & \overset{I_{l,\bar\sigma}}{\nearrow} & \uparrow \rho \\ H_\sigma \otimes_{\mathbb{Q}} \mathbb{Q}_l & & \\ & \underset{I_{l,\bar\sigma\rho}}{\searrow} & \\ & & H_l \end{array}$$

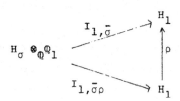

commutes.

The $I_{\infty,\sigma}$ and $I_{l,\bar{\sigma}}$ are called the comparison isomorphisms.

A morphism $f: H \to H'$ of mixed realizations is a family

$$(f_{DR}, f_l, f_\sigma)_{\substack{l \text{ prime number} \\ \sigma:\, k \hookrightarrow \mathbb{C}}}$$

where

1.) $f_{DR} : H_{DR} \to H'_{DR}$ is k-linear and of degree zero for the filtrations W and F .

2.) $f_l : H_l \to H'_l$ is a $\mathbb{Q}_l$-linear G_k-morphism which respects the weight filtrations.

3.) $f_\sigma : H_\sigma \to H'_\sigma$ is a morphism of mixed $\mathbb{Q}$-Hodge structures, i.e., compatible with the filtrations W and F .

4.) f_{DR}, f_l and f_σ correspond under the comparison isomorphisms.

2.2. Remark Assuming 3.) and 4.) , we only have to require

1.)' f_{DR} is k-linear.

2.)' f_l is a $\mathbb{Q}_l$-linear G_k-morphism.

This follows from the properties of the comparison isomorphisms.

2.3. Proposition $\underline{MR}_k$ is an abelian category.

Proof This is clear from the remark above and the fact that mixed Hodge structures form an abelian category [D4]. In particular the morphism f_{DR} , f_l and f_σ are strictly compatible [D4](1.1.5) with the filtrations W and F , and kernels and cokernels are the obvious (componentwise) ones with the induced filtrations and comparison isomorphisms.

2.4. Remark Of course we could separately define categories of de Rham realizations H_{DR} and l-adic realizations H_l and then combine these with the category of mixed $\mathbb{Q}$-Hodge structures (containing the Hodge realizations H_σ). Note however that in general categories of vector spaces with filtrations do not form abelian categories.

2.5. If H is a mixed realization, we define the subobject $W_mH \in \underline{MR}_k$ by

$$W_mH = (W_mH_{DR}, W_mH_l, W_mH_\sigma; I_{\infty,\sigma}|_{W_m}, I_{l,\bar\sigma}|_{W_m})_{l,\sigma,\bar\sigma}$$

where $I_{\infty,\sigma}|_{W_m}$ stands for the restriction

$$W_mH_\sigma \otimes_{\mathbb{Q}} \mathbb{C} \xrightarrow{\sim} W_mH_{DR} \otimes_{k,\sigma} \mathbb{C}$$

of $I_{\infty,\sigma}$ and $I_{l,\bar\sigma}|_{W_m}$ is the restriction

$$W_mH_\sigma \otimes_{\mathbb{Q}} \mathbb{Q}_l \xrightarrow{\sim} W_mH_l$$

of $I_{l,\bar\sigma}$.

2.6. Definition i) A mixed realization H is pure of weight m, if $W_mH = H$ and $W_{m-1}H = 0$.

ii) The category of realizations $\underline{R}_k$ is the full subcategory of $\underline{MR}_k$, whose objects are direct sums of pure realizations.

We see that in general any object H in $\underline{MR}_k$ is a successive extension of the pure realizations $Gr_m^W H := W_mH/W_{m-1}H$.

2.7. There is a natural tensor law on $\underline{MR}_k$ by defining

$$H \otimes H' = (H_{DR} \otimes_k H'_{DR}, H_l \otimes_{\mathbb{Q}_l} H'_l, H_\sigma \otimes_{\mathbb{Q}} H'_\sigma; I_{\infty,\sigma} \otimes I'_{\infty,\sigma}, I_{l,\bar\sigma} \otimes I'_{l,\bar\sigma})_{l,\sigma,\bar\sigma}$$

and taking the natural structures on the components, namely

the induced filtrations (cf. [D4] 1.1.12)

$$W_m(H_{DR} \otimes H'_{DR}) = \sum_{r+s=m} W_r H_{DR} \otimes W_s H'_{DR}$$

$$F^n(H_{DR} \otimes H'_{DR}) = \sum_{p+q=n} F^p H_{DR} \otimes F^q H'_{DR}$$

and similarly for the other realizations, and $\rho \in G_k$ acting by $\rho(x \otimes x') = \rho x \otimes \rho x'$ on $H_1 \otimes H'_1$. Furthermore, by taking the natural commutativity and associativity constraints for vector spaces it is clear, that $\underline{MR}_k$ gets the structure of a tensor category, cf. [DMOS] II 1.1, 1.2, with identity object

$$1 = (k, \mathbb{Q}_1, \mathbb{Q} ; id_{\infty,\sigma} , id_{1,\bar{\sigma}})$$

pure of weight zero ($F^0k = k$, $F^1k = 0$, trivial action of G_k on $\mathbb{Q}_1$, $\mathbb{Q}$ the unique $\mathbb{Q}$-Hodge structure of type $(0,0)$, the comparison isomorphisms induced by $\mathbb{Q} \hookrightarrow \mathbb{C} \overset{\sigma}{\hookleftarrow} k$ and $\mathbb{Q} \hookrightarrow \mathbb{Q}_1 = \mathbb{Q}_1$).

<u>2.8. Definition</u> For $H, H' \in \underline{MR}_k$ define $\underline{Hom}(H,H') \in \underline{MR}_k$ (the "internal Hom") by

a) $H_{DR}(\underline{Hom}(H,H')) = Hom_k(H_{DR},H'_{DR})$ with
$F^n Hom_k(H_{DR},H'_{DR}) = \{f \mid f(F^p H_{DR}) \subseteq F^{p+n}H'_{DR}$ for all $p\}$,
$W_m Hom_k(H_{DR},H'_{DR}) = \{f \mid f(W_r H_{DR}) \subseteq W_{r+m}H'_{DR}$ for all $r\}$,

b) $H_1(\underline{Hom}(H,H')) = Hom_{\mathbb{Q}_1}(H_1,H'_1)$ with G_k-action $(\rho f)(h) = \rho f(\rho^{-1}h)$ for $\rho \in G_k$ and $h \in H_1$, and similar weight filtration,

c) $H_\sigma(\underline{Hom}(H,H')) = Hom_{\mathbb{Q}}(H_\sigma,H'_\sigma)$ with the induced mixed $\mathbb{Q}$-Hodge structure, i.e., with Hodge and weight filtration like above, cf. [D4],

d) the obvious comparison isomorphisms induced by the ones of H and H'.

Then we have a natural (functorial) isomorphism

(2.9) $\underline{\mathrm{Hom}}(T, \underline{\mathrm{Hom}}(H,H')) \xrightarrow{\sim} \underline{\mathrm{Hom}}(T \otimes H, H')$.

<u>2.10.</u> For $H \in \underline{MR}_k$ define the set of <u>absolute Hodge cycles</u> of H by

$$\Gamma(H) = \{(x_{DR}, x_1, x_\sigma)_{1,\sigma} \in H_{DR} \times \prod_1 H_1 \times \prod_\sigma H_\sigma \mid I_{\infty,\sigma}(x_\sigma) = x_{DR}$$
and $I_{1,\bar\sigma}(x_\sigma) = x_1$ for all $\sigma: k \hookrightarrow \mathbb{C}$ and $\bar\sigma: \bar k \hookrightarrow \mathbb{C}$ restricting to σ, $x_{DR} \in F^0 H_{DR} \cap W_0 H_{DR}\}$.

This is a finite-dimensional $\mathbb{Q}$-vector space, as one sees by projection to one H_σ. Note that by properties d) and e) of the comparison isomorphisms $x_1 \in H_1^{G_k} \cap W_0 H_1$ and $x_\sigma \in F^0(H_\sigma \otimes \mathbb{C}) \cap W_0 H_\sigma$ for $x \in \Gamma(H)$. From the definition of $\underline{\mathrm{Hom}}(H,H')$ we see

(2.11) $\mathrm{Hom}(H,H') = \Gamma(\underline{\mathrm{Hom}}(H,H'))$.

In particular, (2.9) implies functorial isomorphisms

(2.12) $\mathrm{Hom}(T, \underline{\mathrm{Hom}}(H,H')) \xrightarrow{\sim} \mathrm{Hom}(T \otimes H, H')$

for $T,H,H' \in \underline{MR}_k$, i.e., the contravariant functor $T \rightsquigarrow \mathrm{Hom}(T \otimes H, H')$ is represented by $\underline{\mathrm{Hom}}(H,H')$. With this we easily obtain

<u>2.13. Theorem</u> $\underline{MR}_k$ is a neutral Tannakian category over $\mathbb{Q}$ (see [SR] and [DMOS] II 2.19), namely a rigid abelian $\mathbb{Q}$-linear tensor category with exact faithful $\mathbb{Q}$-linear tensor functors (<u>fibre functors</u>)

$$H_\sigma : \underline{MR}_k \to \underline{Vec}_{\mathbb{Q}} = \text{category of finite-dimensional } \mathbb{Q}\text{-vector spaces}$$

$$H \mapsto H_\sigma$$

for each $\sigma: k \hookrightarrow \mathbb{C}$.

The proof is routine and rather straightforward, as the axioms of a Tannakian are modeled after the properties of vector spaces and we are dealing with vector spaces (with some additional structure), where all functorial maps are the obvious ones. We only note that the dual $H^\vee$ of $H \in \underline{MR}_k$ is given by

$$\begin{aligned} H^\vee &= \underline{\mathrm{Hom}}\,(H,1) \\ &= (\mathrm{Hom}_k(H_{DR},k), \mathrm{Hom}_{\mathbb{Q}_1}(H_1,\mathbb{Q}_1), \mathrm{Hom}_{\mathbb{Q}}(H_\sigma,\mathbb{Q}) ; (I^{-1}_{\infty,\sigma})^\vee, (I^{-1}_{1,\bar\sigma})^\vee)_{1,\sigma,\bar\sigma} \end{aligned}$$

where $(I^{-1}_{\infty,\sigma})^\vee$ and $(I^{-1}_{1,\bar\sigma})^\vee$ are the transposes of $I^{-1}_{\infty,\sigma}$ and $I^{-1}_{1,\bar\sigma}$, respectively. H_σ is exact, as kernels and cokernels are taken componentwise, and faithful, as $f \in \mathrm{Hom}(H,H')$ is completely determined by $f_\sigma : H_\sigma \to H'_\sigma$.

<u>2.14. Remark</u> There is a canonical isomorphism $\mathrm{Hom}(1,H) \xrightarrow{\sim} \Gamma(H)$ by $f \mapsto (f_{DR}(1), f_1(1), f_\sigma(1))_{1,\sigma}$, in particular $\Gamma(H) \neq 0$ if and only if H contains the object 1 .

<u>2.15. Proposition</u> $\underline{R}_k$ is a (neutral) Tannakian subcategory of $\underline{MR}_k$, which is closed under the formation of subquotients.

<u>Proof</u> Obviously $\underline{R}_k$ is closed under the formation of tensor products and duals, and it contains the identity object 1 . The exactness of the inclusion functor follows from the second statement, which we only have to prove for quotients, as the case of subobjects follows by applying the exact functor $H \mapsto H^\vee$ (twice). But quotients of pure objects are obviously

pure, and the general case follows by induction, as for a quotient $H \oplus H' \twoheadrightarrow H''$ of realizations $H, H' \in \underline{R}_k$ with different weights we must have $\operatorname{Im} H \cap \operatorname{Im} H' = 0$.

<u>2.16. Definition</u> There are natural base extension and restriction functors:

i) For an extension $k \hookrightarrow k'$ of fields as above and $H \in \underline{MR}_k$ define the <u>base extension</u> $H' = H \times_k k' \in \underline{MR}_{k'}$ by

a) $H'_{DR} = H_{DR} \otimes_k k'$ with the induced filtrations,

b) $H'_l = \operatorname{Res}_{G_{k'}} H_l$ via the map $G_{k'} \to G_k$ given by an inclusion $\bar{k} \hookrightarrow \overline{k'}$ of the algebraic closures extending $k \hookrightarrow k'$ (well defined up to conjugacy in G_k) , with the same weight filtration.

c) $H'_{\sigma'} = H_\sigma$ for $\sigma' : k' \hookrightarrow \mathbb{C}$ and $\sigma = \sigma'|_k$, with the same mixed $\mathbb{Q}$-Hodge structure,

d) $I'_{\infty,\sigma'} = I_{\infty,\sigma}$ for σ' and σ as above, via the canonical isomorphism $H'_{DR} \otimes_{k',\sigma'} \mathbb{C} \cong H_{DR} \otimes_{k,\sigma} \mathbb{C}$,

e) $I'_{l,\overline{\sigma'}} = I_{l,\bar{\sigma}}$ for $\overline{\sigma'} : \overline{k'} \hookrightarrow \mathbb{C}$ and $\bar{\sigma} = \overline{\sigma'}|_{\bar{k}}$.

ii) For a finite extension k'/k define the <u>restriction</u> $H = R_{k'/k} H'$ of $H' \in \underline{MR}_k$ by :

a) $H_{DR} = H'_{DR}$ (restriction of scalars to k) with the same filtrations,

b) $H_l = \operatorname{Ind}_{G_{k'}}^{G_k} H'_l$, i.e., the representation induced from $G_{k'}$ to G_k (we may assume $\bar{k} = \overline{k'}$ and $G_{k'} \subseteq G_k$) , with the weight filtration $\operatorname{Ind}_{G_{k'}}^{G_k} W_m H'_l$.

c) $H_\sigma = \bigoplus_{\tau \in J\sigma} H'_\tau$ for $\sigma : k \hookrightarrow \mathbb{C}$, with $J_\sigma = \{\tau : k' \hookrightarrow \mathbb{C} \mid \tau|_k = \sigma\}$ (direct sum of the mixed $\mathbb{Q}$-Hodge structures),

d) $I_{\infty,\sigma}$ for $\sigma : k \hookrightarrow \mathbb{C}$ is given by

$$\bigoplus_{\tau\in J_\sigma} H'_\tau \otimes_{\mathbb{Q}} \mathbb{C} \xrightarrow[\sim]{\oplus I'_{\infty,\tau}} \bigoplus_{\tau\in J_\sigma} H'_{DR} \otimes_{k',\tau} \mathbb{C} \xrightarrow{\sim} H'_{DR} \otimes_{k,\sigma} \mathbb{C}$$

via the canonical isomorphism $k' \otimes_{k,\sigma} \mathbb{C} \xrightarrow{\sim} \bigoplus_{\tau\in J_\sigma} k' \otimes_{k',\tau} \mathbb{C}$,

e) $I_{1,\bar\sigma} : \bigoplus_{\tau\in J_\sigma} H'_\tau \otimes_{\mathbb{Q}} \mathbb{Q}_1 \underset{\sim}{\rightarrow} \operatorname{Ind}_{G_{k'}}^{G_k} H'_1 = \{f: G_k \to H'_1 \mid f(\rho'\rho) = \rho' f(\rho)$ for all $\rho' \in G_{k'}\}$ (with the G_k-action $(\rho f)(\rho_o) = f(\rho_o \rho)$) for $\bar\sigma : \bar k = \overline{k'} \hookrightarrow \mathbb{C}$ with $\bar\sigma|_k = \sigma$ and $\bar\sigma|_{k'} = \sigma'$ is given by

$$a = (a_\tau)_{\tau\in J\sigma} \mapsto f_a : f_a(\rho) = I'_{1,\bar\sigma\rho^{-1}}(a_{\sigma'\rho^{-1}}),$$

which lies in $\operatorname{Ind}_{G_{k'}}^{G_k} H'_1$ by property e) of the comparison isomorphism.

2.17. Definition i) The Tate realization

$$\mathbb{1}(1) = (k(1), \mathbb{Q}_1(1), \mathbb{Q}(1), id_{\infty,\sigma}(1), id_{1,\bar\sigma}(1))_{1,\sigma,\bar\sigma}$$

is pure of weight -2 with:

a) $k(1) = k$ with Hodge filtration $F^o k(1) = 0, F^{-1} k(1) = k(1)$,

b) $\mathbb{Q}_1(1) = \mathbb{Z}_1(1) \otimes_{\mathbb{Z}_1} \mathbb{Q}_1$, where $\mathbb{Z}_1(1) = \varprojlim_n \mu_{1^n}$ and μ_{1^n} is the group of 1^n-th roots of unity in $\bar k$ with the natural G_k-action,

c) $\mathbb{Q}(1)$ is the $\mathbb{Q}$-Hodge structure $2\pi i\, \mathbb{Q}$, of Hodge type $(-1,-1)$ (for any $\sigma: k \hookrightarrow \mathbb{C}$),

d) $id_{\infty,\sigma}(1)$ is induced by $2\pi i\, \mathbb{Q} \hookrightarrow \mathbb{C} \overset{\sigma}{\hookleftarrow} k$,

e) $id_{1,\bar\sigma}(1)$ for $\bar\sigma : \bar k \hookrightarrow \mathbb{C}$ is induced by $2\pi i\, \mathbb{Z} \to \mathbb{Z}_1(1)$, $2\pi i t \mapsto (\bar\sigma^{-1}(\exp \frac{2\pi i t}{1^n}))_n$.

ii) For any mixed realization H and $n \in \mathbb{Z}$ we let

$$H(n) = \begin{cases} H \otimes \mathbb{1}(1)^{\otimes n} & n \geq 0 \\ \underline{\operatorname{Hom}}(\mathbb{1}(1)^{\otimes -n}, H) & n < 0, \end{cases}$$

the n-fold Tate twist of H.

2.18. Remarks i) Twisting shifts the weight by -2 so that $W_m(H(n)) = (W_{m+2n}H)(n)$.

ii) The above definition is compatible with the usual notion of Tate objects and Tate twists in the category of l-adic representations of G_k and the category of Hodge structures, see [D3]. We also obtain a notion of Tate twists in the category of de Rham realizations: $H_{DR}(n)$ equals H_{DR} as k-vector space whereas the Hodge filtration is $F^pH_{DR}(n) = F^{p+n}H_{DR}$. With this we can write $H(n) = (H_{DR}(n), H_l(n), H_\sigma(n); I_{\infty,\sigma}(n), I_{l,\bar{\sigma}}(n))$.

The next two sections are very technical and not needed in the sequel except partially in the proof of 4.7 e), so the reader is advised to skip 2.19 and 2.20 on a first reading.

2.19. $\mathrm{Aut}_k(k')$ acts on $\Gamma(H \times_k k')$ for $H \in \underline{MR}_k$ by $\rho(x_{DR}, x_l, x_\sigma) = (\rho x_{DR}, \rho x_l, x_{\sigma\rho})$ for $\rho \in \mathrm{Aut}_k(k')$, where ρ acts on $H_{DR} \otimes_k k'$ via k' and on H_l via lifting to $\bar{\rho} \in \mathrm{Aut}_k(\overline{k'})$ and the projection $\mathrm{Aut}_k(\overline{k'}) \to G_k$; note that $\mathrm{Gal}(\overline{k'}/k') = G_{k'}$ acts trivially on x_l for $x = (x_{DR}, x_l, x_\sigma) \in \Gamma(H \times_k k')$. If $k' = \bar{k}$ is an algebraic closure of k, then the action of $\mathrm{Gal}(\bar{k}/k) = G_k$ on $\Gamma(H \times_k \bar{k})$ factorizes through a finite quotient, as the map $\Gamma(H \times_k \bar{k}) \hookrightarrow H_{DR} \otimes_k \bar{k}$ is injective and the image of a basis of $\Gamma(H \times_k \bar{k})$ lies in $H_{DR} \otimes_k k'$ for some finite extension k'/k. Obviously $\Gamma(H) \xrightarrow{\sim} \Gamma(H \times_k \bar{k})^{G_k}$.

The above applies in particular to $\mathrm{Hom}(H \times_k k', K \times_k k') = \Gamma(\underline{\mathrm{Hom}}(H \times_k k', K \times_k k')) = \Gamma(\underline{\mathrm{Hom}}(H,K) \times_k k')$ for $H, K \in \underline{MR}_k$. Here $\rho f = (\rho f_{DR}, \rho f_l, f_{\sigma\rho})$ for $\rho \in \mathrm{Aut}_k(k')$ and $f = (f_{DR}, f_l, f_\sigma) \in \mathrm{Hom}(H \times_k k', K \times_k k')$, with $(\rho f_l)(x) = \rho f_l(\rho^{-1}x)$ for $x \in H_l$ and similar for f_{DR}, and one has $\mathrm{Hom}(H,K) \xrightarrow{\sim} \mathrm{Hom}_{G_k}(H \times_k \bar{k}, k \times_k \bar{k})$.

2.20. Proposition Let k' be a finite extension of k .

a) The base extension functor $H \rightsquigarrow H\times_k k'$ is left and right adjoint to the restriction functor $H' \rightsquigarrow R_{k'/k}H'$.

b) There are canonical functorial isomorphisms

$$R_{k'/k}(H')^{\vee} \cong (R_{k'/k}H')^{\vee} \quad , \quad H^{\vee}\times_k k' \cong (H\times_k k')^{\vee}$$

c) There is a canonical functorial isomorphism

$$\Gamma(R_{k'/k}H') \cong \Gamma(H')$$

for $H' \in \underline{MR}_{k'}$.

d) There are functorial isomorphisms, $H \in \underline{MR}_k$ and $H' \in \underline{MR}_{k'}$:

i) $$H \otimes R_{k'/k}H' \cong R_{k'/k}(H\times_k k' \otimes H')$$

ii) $$\underline{\mathrm{Hom}}(H, R_{k'/k}H') \cong R_{k'/k}\,\underline{\mathrm{Hom}}(H\times_k k', H')$$

iii) $$\underline{\mathrm{Hom}}(R_{k'/k}H', H) \cong R_{k'/k}\,\underline{\mathrm{Hom}}(H', H\times_k k') .$$

These isomorphisms - especially the first - will be called the <u>projection formulas</u>; note that we obtain the familiar form $H \otimes \varphi_*H' \cong \varphi_*(\varphi^*H \otimes H')$ if we write $R_{k'/k} = \varphi_*$ and $\times_k k' = \varphi^*$ for φ: Spec $k' \to$ Spec k .

e) For the composition of the maps given by the left and right adjointness in a) we have for $H \in \underline{MR}_k$ and $H' \in \underline{MR}_{k'}$:

$H \xrightarrow{i} R_{k'/k}(H\times_k k') \xrightarrow{q} H$ is multiplication by $[k' : k]$,

$H' \xrightarrow{i} (R_{k'/k}H')\times_k k' \xrightarrow{p} H'$ is the identity.

In particular, H is canonically a direct factor of $R_{k'/k}(H\times_k k')$, and H' is canonically a direct factor of $(R_{k'/k}H')\times_k k'$.

f) There is a natural action of $\mathrm{Aut}_k(k')$ on $R_{k'/k}(H\times_k k')$.

If k'/k is a Galois extension, the map

$$R_{k'/k}(H\times_k k') \xrightarrow{q} H \xrightarrow{i} R_{k'/k}(H\times_k k')$$

is the trace under $\mathrm{Aut}_k(k') = \mathrm{Gal}(k'/k)$.

<u>Proof</u> The maps

(2.20.1) $$H \underset{q}{\overset{i}{\rightleftarrows}} R_{k'/k}(H\times_k k') \qquad H' \underset{j}{\overset{p}{\leftrightarrows}} (R_{k'/k}H')\times_k k'$$

such that (i,p) give the left adjointness and (j,q) the right adjointness of $\times_k k'$ to $R_{k'/k}$ are defined as follows:

$$H_{DR} \underset{q_{DR}}{\overset{i_{DR}}{\rightleftarrows}} H_{DR} \otimes_k k' \qquad H'_{DR} \underset{j_{DR}}{\overset{p_{DR}}{\leftrightarrows}} H'_{DR} \otimes_k k'$$

$$h \longmapsto h \otimes 1 \qquad a'\cdot h' \longleftarrow h' \otimes a'$$

$$h\cdot tr_{k'/k}a \longleftarrow h \otimes a \qquad h' \longmapsto (h' \otimes 1)_o$$

where $(h' \otimes 1)_o$ denotes the projection of $h' \otimes 1$ to the subspace of $H'_{DR} \otimes_k k'$ on which both k'-structures coincide - this is the projection by the idempotent of $k' \otimes_k k'$ which gives a section to the multiplication morphism $k' \otimes_k k' \to k'$.

$$H_1 \underset{q_1}{\overset{i_1}{\rightleftarrows}} \mathrm{Ind}_{G_{k'}}^{G_k} H_1 \qquad H'_1 \underset{j_1}{\overset{p_1}{\leftrightarrows}} \mathrm{Ind}_{G_{k'}}^{G_k} H'_1$$

$$h \longmapsto f:\ \rho \mapsto \rho h \qquad f(1) \longleftarrow f:\ G_k \to H'_1$$

$$\sum_{\rho\in G_k/G_{k'}} \rho f(\rho^{-1}) \longleftarrow f:\ G_k \to H_1 \qquad h' \longmapsto f:\ f(\rho) = \begin{cases} \rho h', & \rho\in G_{k'} \\ 0, & \rho\notin G_{k'} \end{cases}$$

where the functor $\mathrm{Ind}_{G_{k'}}^{G_k}$ is defined as in 2.16 ii)e), and $\rho \in G_k/G_{k'}$ means choosing any system of representatives in G_k for the cosets $\rho \cdot G_{k'}$ (the sum is independent of the choice). Finally, for $\sigma': k' \hookrightarrow \mathbb{C}$, $\sigma = \sigma'|_k$ and $J_\sigma = \{\tau: k' \hookrightarrow \mathbb{C} \mid \tau|_k = \sigma\}$ we define

$$H_\sigma \underset{q_\sigma}{\overset{i_\sigma}{\rightleftarrows}} \bigoplus_{\tau\in J_\sigma} H_\sigma \qquad H'_{\sigma'} \underset{j_{\sigma'}}{\overset{p_{\sigma'}}{\leftrightarrows}} \bigoplus_{\tau\in J_\sigma} H'_\iota$$

$$h \longmapsto (h)_{\tau\in J_\sigma} \qquad h'_{\sigma'} \longleftarrow (h'_\tau)_{\tau\in J_\sigma}$$

$$\sum_{\tau\in J_\sigma} h_\tau \longleftarrow (h_\tau)_{\tau\in J_\sigma} \qquad h' \longmapsto (h'_\tau)_{\tau\in J_\sigma}, h'_{\sigma'}=h',\ h'_\tau = 0 \text{ for } \tau\neq\sigma'.$$

It is lengthy but easy to check that these maps are compatible with the comparison isomorphisms, which are given as follows. For $R_{k'/k}(H \times_k k')$, by 2.16 $I_{\infty,\sigma}$ is given by

$$\begin{array}{ccccc} (\bigoplus_{\tau\in J_\sigma} H_\sigma) \otimes \mathbb{C} & \longrightarrow & (H_{DR} \otimes_k k')\otimes_{k,\sigma}\mathbb{C} & & h\otimes a\otimes c \\ \| & & \downarrow \wr & & \downarrow \\ \bigoplus_{\tau\in J_\sigma} (H_\sigma \otimes \mathbb{C}) & \xrightarrow{\bigoplus_{\tau\in J_\sigma} I_{\infty,\sigma}} & \bigoplus_{\tau\in J_\sigma} H_{DR} \otimes_{k,\sigma} \mathbb{C} & & (h\otimes\tau(a)\cdot c)_\tau \ , \end{array}$$

and $I_{1,\bar\sigma}$ for $\bar\sigma : \bar k \hookrightarrow \mathbb{C}$ with $\sigma' = \bar\sigma|_{k'}$ by

$$\begin{array}{ccc} \bigoplus_{\tau\in J_\sigma} (H_\sigma \otimes \mathbb{Q}_l) & \to & \mathrm{Ind}_{G_{k'}}^{G_k} H_l \\ a = (a_\tau)_\tau & \mapsto & f_a : G_k \to H_l \\ & & f_a(\rho) = I_{l,\bar\sigma\rho^{-1}}(a_{\sigma'\rho^{-1}}) \quad . \end{array}$$

For $H'' = (R_{k'/k}H')\times_k k'$, $I''_{\infty,\sigma'}$ is determined by the commutative diagram

$$\begin{array}{ccccc} (\bigoplus_{\tau\in J_\sigma} H'_\tau) \otimes \mathbb{C} & \xrightarrow{I''_{\infty,\sigma'}} & (H'_{DR}\otimes_k k') \otimes_{k',\sigma'}\mathbb{C} & , & h' \otimes a' \otimes c \\ \| & & \downarrow \wr & & \downarrow \\ (\bigoplus_{\tau\in J_\sigma} H'_\tau) \otimes \mathbb{C} & \xrightarrow{I_{\infty,\sigma}} & H'_{DR} \otimes_{k,\sigma}\mathbb{C} & & h' \otimes \sigma'(a')c \\ \| & & \downarrow \wr & & \downarrow \\ \bigoplus_{\tau\in J_\sigma} (H'_\tau \otimes \mathbb{C}) & \xrightarrow{\bigoplus_{\tau\in J_\sigma} I'_{\infty,\tau}} & \bigoplus_{\tau\in J_\sigma} H'_{DR}\otimes_{k',\tau}\mathbb{C} & & (h' \otimes \sigma'(a')c)_\tau \ , \end{array}$$

where on the right side $(H'_{DR} \otimes_k k')_o \otimes_{k',\sigma'}\mathbb{C}$ is mapped isomorphically onto the factor $H'_{DR} \otimes_{k',\sigma'}\mathbb{C}$, while $I''_{l,\bar\sigma}$ is

$$\bigoplus_{\tau\in J_\sigma} (H'_\tau \otimes \mathbb{Q}_l) \quad \to \quad \mathrm{Ind}_{G_{k'}}^{G_k} H'_l$$

$$a = (a'_\tau)_\tau \qquad \mapsto \qquad f_a : G_k \to H'_1$$
$$f_a(\rho) = I'_{1,\bar{\sigma}\rho^{-1}}(a_{\sigma'\rho^{-1}}) ,$$

with the same notations as above.

Finally it is clear from the definitions (especially using 2.2) that the above maps respect all relevant structures like filtrations, Galois action etc. . We now prove the claims of the proposition.

a) Once i, j, p and q are proved to define morphisms of mixed realizations, the defining properties of adjunction morphisms may be checked in anyone of the realizations. For example, the fact that the compositions

$$H \times_k k' \xrightarrow{i_{H\times_k k'}} (R_{k'/k}(H\times_k k'))\times_k k' \xrightarrow{p_{H\times_k k'}} H \times_k k'$$

$$R_{k'/k} \xrightarrow{i_{R_{k'/k}H'}} R_{k'/k}((R_{k'/k}H')\times_k k') \xrightarrow{R_{k'/k}p_{H'}} R_{k'/k}H'$$

give the identities, is easily seen in the de Rham realization:

$$\begin{array}{ccccc} H_{DR} \otimes_k k' & \longrightarrow & (H_{DR} \otimes_k k') \otimes_k k' & \longrightarrow & H_{DR} \otimes_k k' \\ h \otimes a' & \longmapsto & (h \otimes 1) \otimes a' & \longmapsto & (h \otimes 1)\cdot a' = h \otimes a' \end{array}$$

and

$$\begin{array}{ccccc} H'_{DR} & \longrightarrow & H'_{DR} \otimes_k k' & \longrightarrow & H'_{DR} \\ h' & \longmapsto & h' \otimes 1 & \longmapsto & h'\cdot 1 = h' . \end{array}$$

That the following compositions give the identities

$$R_{k'/k}H' \xrightarrow{R_{k'/k}j_{H'}} R_{k'/k}((R_{k'/k}H')\times_k k') \xrightarrow{q_{R_{k'/k}H'}} R_{k'/k}H'$$

$$H \times_k k' \xrightarrow{j_{H\times_k k'}} (R_{k'/k}(H\times_k k'))\times_k k' \xrightarrow{q_{H\times_k k'}} H \times_k k' ,$$

can be checked in the Hodge realizations:

$$\bigoplus_{\tau\in J_\sigma} H'_\tau \xrightarrow{\oplus j_\tau} \bigoplus_{\tau\in J_\sigma} \left(\bigoplus_{\rho\in J_\sigma} H'_\rho \right) \xrightarrow{\Sigma} \oplus H'_\tau$$

$$(h'_\tau)_\tau \longmapsto ((h'_\tau \cdot \delta_{\tau\rho})_\rho)_\tau \longmapsto \sum_{\tau \in J_\sigma} (h'_\tau \delta_{\tau\rho})_\rho = (h'_\tau)_\tau$$

$$H_\sigma \xrightarrow{j_{\sigma'}} \bigoplus_{\tau \in J_\sigma} H_\sigma \xrightarrow{q_{\sigma'}} H_\sigma$$

$$h \longmapsto (h \cdot \delta_{\tau\sigma'})_\tau \longmapsto \sum_\tau h\delta_{\tau\sigma'} = h \; .$$

b) The first isomorphism is the special case $H = 1_k$ in d) iii), as $1_{k'} = 1_k \times_k k'$. The second follows from the more general isomorphism

(2.20.2) $\quad \underline{\mathrm{Hom}}(H,K) \times_k k' \xrightarrow{\sim} \underline{\mathrm{Hom}}(H \times_k k', K \times_k k')$

for $H,K \in \underline{M}_k$ which is obvious and also a formal consequence (see [DMOS]II 1.9) of the fact that $H \rightsquigarrow H \times_k k'$ is a tensor functor, by the obvious isomorphism

(2.20.3) $\quad (H \otimes K) \times_k k' \cong (H \times_k k') \otimes (K \times_k k')$

which is given by

$$(H_{DR} \otimes_k K_{DR}) \otimes_k k' \xleftarrow{\sim} (H_{DR} \otimes_k k') \otimes_{k'} (K_{DR} \otimes_k k')$$

$$(x \otimes y) \otimes a' \cdot b' \longleftarrow (x \otimes a') \otimes (y \otimes b')$$

and the identity in the other realizations.

c) By 2.14, this is a special case of the adjunction; the map is given by

$$(x'_{DR}, x'_1, x_{\sigma'})_{1,\sigma'} \mapsto (x'_{DR}, \underline{x'_1}, (x'_\tau)_{\tau \in J_\sigma})_{1,\sigma} \quad .$$

d) It is easily checked that

$$H \otimes \varphi_* H' \xrightarrow{i} \varphi_* \varphi^* (H \otimes \varphi_* H') \xrightarrow[\sim]{(2.20.3)} \varphi_* (\varphi^* H \otimes \varphi^* \varphi_* H') \xrightarrow{p} \varphi_* (\varphi^* H \otimes H')$$

is an isomorphism; for example

$$H_{DR} \otimes_k H'_{DR} \to (H_{DR} \otimes_k H'_{DR}) \otimes_k k' \xrightarrow{\sim} (H_{DR} \otimes_k k') \otimes_{k'} (H'_{DR} \otimes_k k') \to (H_{DR} \otimes_k k') \otimes_{k'} H'_{DR}$$

$$h \otimes h' \mapsto (h \otimes h') \otimes 1 \mapsto (h \otimes 1) \otimes (h' \otimes 1) \mapsto (h \otimes 1) \otimes h'$$

is obviously an isomorphism. Writing $R = R_{k'/k}$ we have

$$\begin{array}{ccc}
\mathrm{Hom}(T,\underline{\mathrm{Hom}}(H,RH')) & & \mathrm{Hom}(T,R\ \underline{\mathrm{Hom}}(H\times k',H')) \\
\wr\Vert & & \wr\Vert \text{adjunction} \\
\mathrm{Hom}(T\otimes H,RH') & & \mathrm{Hom}(T\times k',\underline{\mathrm{Hom}}(H\times k',H')) \\
\wr\Vert \text{adjunction} & & \wr\Vert \\
\mathrm{Hom}((T\otimes H)\times k',H') & \underset{(2.20.3)}{\cong} & \mathrm{Hom}((T\times k')\otimes(H\times k'),H')
\end{array}$$

for any $T \in \underline{MR}_k$, which gives ii). Similarly, we have

$$\begin{array}{ccc}
\mathrm{Hom}(T,\underline{\mathrm{Hom}}(RH',H)) & & \mathrm{Hom}(T,R\ \underline{\mathrm{Hom}}(H',H\times k')) \\
\wr\Vert & & \wr\Vert \text{ adjunction} \\
\mathrm{Hom}(T\otimes RH',H) & & \mathrm{Hom}(T\times k',\underline{\mathrm{Hom}}(H',H\times k')) \\
\wr\Vert\ \text{i)} & & \wr\Vert \\
\mathrm{Hom}(R((T\times k')\otimes H'),H) & \underset{\text{adjunction}}{\overset{\sim}{\longrightarrow}} & \mathrm{Hom}((T\times k')\otimes H', H\times k')
\end{array}$$

for any $T \in \underline{MR}_k$, and therefore iii).

e) The first part is immediately clear from the definition of the maps. For the second note that $\underline{M}_k$ is a $\mathbb{Q}$-linear category and $[k' : k]$ is invertible in $\mathbb{Q}$.

f) Formally, the action of $\tau \in \mathrm{Aut}_k(k')$ on $\varphi_*\varphi^*$ is given by the adjunction $\mathrm{id} \to (\mathrm{Spec}\ \tau)_*(\mathrm{Spec}\ \tau)^*$, inducing $\varphi_*\varphi^* \to \varphi_*(\mathrm{Spec}\ \tau)_*(\mathrm{Spec}\ \tau)^*\varphi^* = \varphi_*\varphi^*$. Explicitely, $\tau : R_{k'/k}(H \times_k k') \to R_{k'/k}(H \times_k k')$ is described by the maps

$$\begin{array}{ccccl}
H_{DR}\otimes_k k' & \to & H_{DR}\otimes_k k' & , & x\otimes a' \mapsto x\otimes\tau(a') \\
\mathrm{Ind}_{G_{k'}}^{G_k} H_1 & \to & \mathrm{Ind}_{G_{k'}}^{G_k} H_1 & , & f \to {}^{\tau}f : {}^{\tau}f(\rho) = \tau f(\tau^{-1}\rho) \\
\bigoplus_{\rho\in J_\sigma} H_\sigma & \to & \bigoplus_{\rho\in J_\sigma} H_\sigma & , & (h_\rho)_\rho \to (h_{\rho\tau})_\rho \ ,
\end{array}$$

where τ also denotes a lifting in G_k. With this, it follows directly from the definitions that $i \circ q = \sum_{\tau\in \mathrm{Gal}(k'/k)} \tau$ for k'/k Galois. q.e.d.

§3. The mixed realization of a smooth variety

Let k, $\bar{k}$ and G_k be as in the previous paragraphs and let $\underline{\overset{o}{V}}_k$ be the category of smooth quasi-projective varieties over k. We want to construct functors $H^n : \underline{\overset{o}{V}}_k \to \underline{MR}_k$ for $n \in \mathbb{Z}$, associating to each $U \in \underline{\overset{o}{V}}_k$ its n-th realization $H^n(U) = (H^n_{DR}(U), H^n_l(U), H^n_\sigma(U); I_{\infty,\sigma}, I_{l,\bar{\sigma}})_{l,\sigma,\bar{\sigma}}$.

<u>3.1.</u> Let $H^n_{DR}(U) = H^n_{DR}(U/k) = \mathbb{H}^n(U_{Zar}, \Omega^\cdot_{U/k})$ (Zariski-hypercohomology of the de Rham complex). For the filtrations F and W we follow Deligne's construction of mixed Hodge structures [D4], we only have to show that everything is defined over k.

<u>3.2.</u> By Hironaka's result on resolution of singularities [Hir], there exists a smooth projective variety X over k and an open immersion $j: U \hookrightarrow X$, such that $Y = X \setminus U$ (with the reduced subscheme structure) is the union of smooth divisors Y_i, $i = 1,\dots,N$, with normal crossings. Recall that this means one of the following equivalent conditions to hold

a) Each $x \in X$ has an affine open neighborhood V such that V is étale over the affine space $\mathbb{A}^d_k$, $d = \dim X$, via "coordinates" $x_1,\dots,x_d \in \Gamma(V, \mathcal{O}_X)$, and $V \cap Y$ is defined by $x_1 \dots x_\nu = 0$ for some $0 \leq \nu \leq \min(d,N)$, i.e., is the pull-back of the union of the ν first coordinate hyperplanes in $\mathbb{A}^d_k$.

b) If f_i is a local equation for Y_i at $x \in X$ and $J_x = \{i \in \{1,\dots,N\}) \mid f_i$ is not a unit at $x\}$, then $(f_i)_{i \in J_x}$ is part of a regular system of parameters at x.

<u>3.3.</u> The sheaf $\Omega^1_X\langle Y\rangle$ of differentials with <u>logarithmic poles along Y</u> is defined as the subsheaf of $j_*\Omega^1_U$ generated over $\mathcal{O}_X$ by Ω^1_X and $d\log j_*\mathcal{O}^\times_U$, where we write Ω^1_X and Ω^1_U for the

sheaves $\Omega^1_{X/k}$ and $\Omega^1_{U/k}$ of relative differentials, and $\mathrm{dlog}\, f = \frac{df}{f}$ for a unit f. Then $\Omega^1_X\langle Y\rangle$ is locally free, namely for an open affine V like in 3.2a), $\Omega^1_X\langle Y\rangle|_V$ is a free $\mathcal{O}_V$-module with basis $\frac{dx_1}{x_1}, \ldots, \frac{dx_\nu}{x_\nu}, x_{\nu+1}, \ldots, x_d$. By defining $\Omega^p_X\langle Y\rangle = \overset{p}{\wedge}\, \Omega^1_X\langle Y\rangle$, and taking the differentials d to be the restrictions of those for $j_*\Omega^\bullet_U$ one obtains the <u>logarithmic de Rham complex</u> $\Omega^\bullet_X\langle Y\rangle$:

$$\mathcal{O}_X \xrightarrow{d} \Omega^1_X\langle Y\rangle \xrightarrow{d} \Omega^2_X\langle Y\rangle \rightarrow \ldots \quad .$$

Its formation is compatible with base change in k and étale base change in X.

<u>3.4. Lemma</u> The map $\Omega^\bullet_X\langle Y\rangle \rightarrow j_*\Omega^\bullet_U$ induces isomorphisms in the Zariski-hypercohomology

$$\mathbb{H}^n(X,\Omega^\bullet_X\langle Y\rangle) \xrightarrow{\sim} \mathbb{H}^n(X,j_*\Omega^\bullet_U) \xrightarrow{\sim} \mathbb{H}^n(U,\Omega^\bullet_U) = H^n_{DR}(U) \ .$$

<u>Proof</u> As Y is a divisor with normal crossing, j is affine, and therefore

$$R^p j_*\Omega^q_U = 0 \quad \text{for } p > 0 \text{ and all } q \ .$$

This implies the second isomorphism.

The first isomorphism follows by base change from the corresponding fact over $\mathbb{C}$, which can be proved by analytic methods, see [D3] II 3.14. One can also give a purely algebraic proof along these lines, by replacing the considerations for a polydisk by similar ones for the affine space A^d_k, using the criterion 3.2 a).

<u>3.5.</u> The <u>weight filtration</u> W on $\Omega^\bullet_X\langle Y\rangle$ is defined by

$$W_m\Omega_X^p\langle Y\rangle = \begin{cases} 0 & , \quad m < 0 , \\ \Omega_X^{p-m}\wedge\Omega_X^m\langle Y\rangle & , \quad 0 \leq m < p \\ \Omega_X^p\langle Y\rangle & , \quad m \geq p . \end{cases}$$

The differentials d respect these subspaces, and so indeed we get an increasing filtration of $\Omega_X^\bullet\langle Y\rangle$ by subcomplexes $W_m\Omega_X^\bullet\langle Y\rangle$ (Note that $W_{-1} = 0$, $W_o = \Omega_X^\bullet$ and $W_d = \Omega_X^\bullet\langle Y\rangle$, when $d = \dim X$) . For an integer $m \geq 0$ and indices $1 \leq i_1 < i_2 < \ldots < i_m \leq N$ consider the map

$$\Omega_X^{p-m} \longrightarrow W_m\Omega_X^p\langle Y\rangle ,$$

which is locally given by

$$\alpha \longmapsto \alpha \wedge \frac{dx_{i_1}}{x_{i_1}} \wedge\ldots\wedge \frac{dx_{i_m}}{x_{i_m}} ,$$

where α is a holomorphic (p-m)-form and x_i is a local equation for Y_i. The induced map

$$\Omega_X^{p-m} \rightarrow Gr_m^W\Omega_X^p\langle Y\rangle$$

does not depend on the choice of the local equations x_i , as for other equations x_i' one has

$$\frac{dx_i}{x_i} - \frac{dx_i'}{x_i'} = \frac{d(\frac{x_i}{x_i'})}{\frac{x_i}{x_i'}} ,$$

which is holomorphic. Also it factorizes through $(b_{i_1\ldots i_m})_*\Omega^{p-m}_{Y_{i_1\ldots i_m}}$, where $b_{i_1\ldots i_m} : Y_{i_1\ldots i_m} = Y_{i_1}\cap\ldots\cap Y_{i_m} \hookrightarrow X$ is the closed immersion, as $\beta \wedge dx_{i_\nu}$ and $x_{i_\nu} \cdot \alpha$ are mapped to zero .
We obtain an induced map

$$\rho_m^p : (i_m)_*\Omega^{p-m}_{Y(m)} \rightarrow Gr_m^W \Omega_X^p\langle Y\rangle ,$$

where $Y^{(m)} = \dot{\bigcup\limits_{1\leq i_1<\ldots<i_m\leq N}} Y_{i_1\ldots i_m}$ is the disjoint union of the m-fold intersections of the $Y_1,\ldots,Y_N$, which is also the

normalization of $Y^m = \bigcup_{1 \leq i_1 < \ldots < i_m \leq N} Y_{i_1 \ldots i_m} \subseteq X$, and where $i_m : Y^{(m)} \to X$ is the canonical map (by definition $Y^{(0)} = X$).

3.6. Lemma The map of complexes

$$(3.6.1) \qquad \rho_m^{\cdot} : (i_m)_* \Omega^{\cdot}_{Y^{(m)}}[-m] \to Gr_m^W \Omega_X^{\cdot}\langle Y\rangle$$

is an isomorphism (recall that $(K^{\cdot}[n])^i = K^{i+n}$ for a complex $K^{\cdot}$).

Proof By criterion 3.2 a) and étale base change this need only be checked for the case $X = A_k^d$ with coordinates $x_1, \ldots x_d$ and Y the union of the hyperplanes $Y_i = \{x_i = 0\}$ for $i = 1, \ldots, \nu$. By using the canonical decomposition of both sides for products $X = X_1 \times X_2$ and $Y = Y_1 \times X_2 \cup X_1 \times Y_2$ (i.e., $U = U_1 \times U_2$) as in [D2] II 3.6 one easily reduces to the case $d = 1$, where in the only interesting case $m = 1$ and $p = 1$ we obviously have an isomorphism

$$(i_1)_* \mathcal{O}_Y \xrightarrow{\sim} \Omega_X^1\langle Y\rangle / \Omega_X^1 \quad ,$$

via the exact sequence

$$0 \to \Omega_X^1 \to \Omega_X^1\langle Y\rangle \to (i_1)_* \mathcal{O}_Y \to 0$$

corresponding to the exact sequence of $k[x]$ -modules

$$0 \to k[x]dx \to k[x]\frac{dx}{x} \to k \to 0 \quad .$$

This proof also works in characteristic p , while for characteristic zero the lemma also follows from the corresponding statement about analytic sheaves proved in [D3] II 3.6, by using base extension to $\mathbb{C}$ and the GAGA principle, as the ρ_m^p are linear.

3.7. Remark If $(\tau_n)_{n \in \mathbb{Z}}$ is the canonical increasing filtration (compare [D4] (1.4.6.)) of a $\mathbb{Z}$-graded complex $K^\bullet$

$$\begin{array}{lcccccccccc}
\tau_n K^\bullet & := & \ldots \to K^{n-1} & \to & \operatorname{Ker} d_n & \to & 0 & \to & 0 & \to & \ldots \\
\cap| & & \| & & \downarrow & & \downarrow & & \downarrow & & \\
K^\bullet & := & \ldots \to K^{n-1} & \xrightarrow{d_{n-1}} & K^n & \xrightarrow{d_n} & K^{n+1} & \to & K^{n+2} & \to & \ldots,
\end{array}$$

we have $\tau_n \Omega^\bullet_X\langle Y\rangle \subseteq W_n \Omega^\bullet_X\langle Y\rangle$:

$$\begin{array}{lcccccccc}
\tau_n \Omega^\bullet_X\langle Y\rangle & : & \ldots \to \Omega^{n-1}_X\langle Y\rangle & \to & \operatorname{Ker} d_n & \to & 0 & \to & \ldots \\
& & \| & & \downarrow & & \downarrow & & \\
W_n \Omega^\bullet_X\langle Y\rangle & : & \ldots \to \Omega^{n-1}_X\langle Y\rangle & \to & \Omega^n_X\langle Y\rangle & \to & \Omega^1_X \wedge \Omega^n_X\langle Y\rangle \to & & \ldots
\end{array}$$

However, in contrast to the analytic case (see [D4] (3.1.8)), the identity map $(\Omega^\bullet_X\langle Y\rangle, \tau) \to (\Omega^\bullet_X\langle Y\rangle, W)$ of the algebraic complexes is <u>not</u> a quasi-isomorphism of filtered complexes in general: $\mathrm{Gr}^\tau_n\, \Omega^\bullet_X\langle Y\rangle = \mathcal{H}^n(\Omega^\bullet_X\langle Y\rangle)$ is concentrated in dimension n which is in general not true for $\mathrm{Gr}^W_n \Omega^\bullet_X\langle Y\rangle \cong (i_n)_* \Omega^\bullet_{Y(n)}[-n]$ (whereas the complex analytic version of the latter <u>is</u> concentrated in dimension n , quasi-isomorphic to $(i_n)_*\, \mathbb{C}_{Y(n)}[-n]$).

<u>3.8.</u> The <u>"stupid" filtration</u> $(\sigma^n)_{n\in\mathbb{Z}}$ of $\Omega^\bullet_X\langle Y\rangle$ is given by

$$\begin{array}{lcccccccc}
\sigma^n \Omega^\bullet_X\langle Y\rangle & : & \ldots \to & 0 & \to & \Omega^n_X\langle Y\rangle & \to & \Omega^{n+1}_X\langle Y\rangle & \to \ldots \\
\cap| & & & \downarrow & & \| & & \| & \\
\Omega^\bullet_X\langle Y\rangle & : & \ldots & \to \Omega^{n-1}_X\langle Y\rangle & \to & \Omega^n_X\langle Y\rangle & \to & \Omega^{n+1}_X\langle Y\rangle & \to \ldots .
\end{array}$$

<u>3.9.</u> Recall that for a decreasing biregular filtration $(F^p)_{p\in\mathbb{Z}}$ of a complex $K^\bullet$ of sheaves on X there is an associated spectral sequence for the hypercohomology ([D 4] (1.4.5.))

$$E_1^{p,q} = \mathbb{H}^{p+q}(X, \mathrm{Gr}^p_F K^\bullet) \Rightarrow \mathbb{H}^{p+q}(X, K^\bullet) ,$$

where the differentials $d_1^{p,q}$ are the connecting morphisms for the short exact sequences

$$0 \to Gr_F^{p+1}K^\cdot \to F^pK^\cdot/F^{p+2}K^\cdot \to Gr_F^pK^\cdot \to 0 .$$

The filtration $(F^p)_{p\in\mathbb{Z}}$ on the limit term $E^n = \mathbb{H}^n(X,K^\cdot)$ for which $Gr_F^pE^n \cong E_\infty^{p,n-p}$, is given by

$$F^p\mathbb{H}^n(X,K^\cdot) = Im(\mathbb{H}^n(X,F^pK^\cdot) \to \mathbb{H}^n(X,K^\cdot)) .$$

We apply this to obtain the weight filtration and the Hodge filtration on $H_{DR}^n(U) = \mathbb{H}^n(X,\Omega_X^\cdot\langle Y\rangle)$.

<u>3.10. Defintion</u> The Hodge filtration F on $H_{DR}^n(U)$ is the filtration induced by the spectral sequence

$$(3.10.1) \qquad {}_FE_1^{p,q} = \mathbb{H}^{p+q}(X,Gr_\sigma^p\Omega_X^\cdot\langle Y\rangle) \Rightarrow \mathbb{H}^{p+q}(X,\Omega_X^\cdot\langle Y\rangle)$$

associated to the filtration $(\sigma^n)_{n\in\mathbb{Z}}$, i.e.,

$$F^p\mathbb{H}^n(X,\Omega_X^\cdot\langle Y\rangle) = Im(\mathbb{H}^n(X,\sigma^p\Omega_X^\cdot\langle Y\rangle) \to \mathbb{H}^n(X,\Omega_X^\cdot\langle Y\rangle)) .$$

By the isomorphism of complexes

$$(3.10.2) \qquad Gr_\sigma^p\Omega_X^\cdot\langle Y\rangle \cong \Omega_X^p\langle Y\rangle[-p]$$

(where a sheaf K is identified with the complex $K^\cdot$ such that $K^0 = K$ and $K^i = 0$ for $i \neq 0$) the spectral sequence can be written as

$$(3.10.3) \qquad {}_FE_1^{p,q} = \mathbb{H}^q(X,\Omega_X^p\langle Y\rangle) \Rightarrow \mathbb{H}^{p+q}(X,\Omega_X^\cdot\langle Y\rangle) .$$

We also can apply 3.9 to the increasing filtration W , by passing to the decreasing filtration $(W^n = W_{-n})_{n\in\mathbb{Z}}$ and translating back, but in addition there is a shift involved.

<u>3.11. Definition</u> The weight filtration W on $H_{DR}^n(U)$ is obtained from the filtration W' induced by the spectral sequence

$$(3.11.1)\quad {}_W\tilde{E}_1^{p,q} = \mathbb{H}^{p+q}(X, Gr^W_{-p}\Omega^\cdot_X\langle Y\rangle) \Rightarrow \mathbb{H}^{p+q}(X,\Omega^\cdot_X\langle Y\rangle)$$

of the filtration W of $\Omega^\cdot_X\langle Y\rangle$ by an n-fold shift: $W = W'[n]$, i.e.,

$$W_{n+k}\mathbb{H}^n(X,\Omega^\cdot_X\langle Y\rangle) = Im(\mathbb{H}^n(X,W_k\Omega^\cdot_X\langle Y\rangle) \to \mathbb{H}^n(X,\Omega^\cdot_X\langle Y\rangle)) .$$

<u>3.12.</u> By the isomorphism (3.6.1)

$$\rho^\cdot_m : Gr^W_m\Omega^\cdot_X\langle Y\rangle \cong (i_m)_*\Omega^\cdot_{Y^{(m)}}[-m]$$

and the fact that i_m is finite and therefore induces an isomorphism in Zariski hypercohomolgy, we have an isomorphism

$$(3.12.1)\quad \mathbb{H}^{p+q}(X,Gr^W_{-p}\Omega^\cdot_X\langle Y\rangle) \cong \mathbb{H}^{2p+q}(Y^{(-p)},\Omega^\cdot_{Y^{(-p)}})$$

(there is a misprint in [D4](3.2.4.1)). On both sides there are natural filtrations F which differ by a shift; in fact the isomorphism ρ'_m is an isomorphism of filtered complexes, if one takes the filtration induced by σ on the left side and the filtration $\sigma(Y^{(m)})[-m]$ obtained by shifting $(\sigma[-m]^i = \sigma^{i-m})$ from the stupid filtration of $\Omega^\cdot_{Y^{(m)}}$. Thus we can rewrite the spectral sequence (3.11.1) as

$$(3.12.2)\quad {}_W\tilde{E}_1^{p,q} = \mathbb{H}^{2p+q}(Y^{(-p)},\Omega^\cdot_{Y^{(-p)}})(p) \Rightarrow \mathbb{H}^{p+q}(X,\Omega^\cdot_X\langle Y\rangle) ,$$

which is compatible with the Hodge filtrations of X and the $Y^{(-p)}$, if (p) indicates the shift of the filtration as in 2.18: $A(m) = A$ as abelian group, with filtration $F^p(A(m) = F^{p+m}A$. It is convenient to renumber ${}_W\tilde{E}_1^{p,q} = {}_WE_2^{2p+q,-p}$ to write the spectral sequence as

$$(3.12.3)\quad {}_WE_2^{p,q} = \mathbb{H}^p(Y^{(q)},\Omega^\cdot_{Y^{(q)}})(-q) \Rightarrow \mathbb{H}^{p+q}(X,\Omega^\cdot_X\langle Y\rangle) .$$

<u>3.13.</u> The filtrations F and W on $H^n_{DR}(U)$ are independent of the particular choice of X and are functorial in U,

as one sees like in [D4] (3.2.11). This gives the de Rham realization of U .

3.14. For the l-adic realization we let

$$H_l^n(U) = H_{et}^n(U \times_k \bar{k}, \mathbb{Q}_l)$$

(étale cohomology), with the action of $\rho \in G_k$ induced via functoriality from $U \times_k \bar{k} \xrightarrow{id \times \rho} U \times_k \bar{k}$. Let W' be the increasing filtration induced by the Leray spectral sequence

$$(3.14.1) \quad E_2^{p,q} = H_{et}^p(X \times_k \bar{k}, R^q \bar{j}_* \mathbb{Q}_l) \Rightarrow H^{p+q}(U \times_k \bar{k}, \mathbb{Q}_l) = E^{p+q}$$

i.e., $0 = W'_{-1}E^n \subseteq W'_0 E^n \subseteq \ldots \subseteq W'_n E^n = E^n$ with $Gr_q^{W'} E^n = E_\infty^{n-q,q}$. Then the weight filtration W of $H_l^n(U)$ is defined as $W'[n]$, i.e., $0 = W_{n-1}E^n \subseteq W_n E^n \subseteq \ldots \subseteq W_{2n}E^n = E^n$ with $Gr_{n+k}^W E^n = E_\infty^{n-k,k}$. It is independent of X and functorial by 3.18 below and the analogous result for $H_\sigma^n(U)$ ([D4](3.2.11)).

3.15. For $\sigma: k \hookrightarrow \mathbb{C}$ we let

$$H_\sigma^n(U) = H^n(U \times_{k,\sigma} \mathbb{C}, \mathbb{Q})$$

(singular cohomology or analytic sheaf cohomology) with the mixed $\mathbb{Q}$-Hodge structure defined by Deligne [D4] for the smooth variety $\sigma U = U \times_{k,\sigma} \mathbb{C}$ over $\mathbb{C}$. If $\sigma X = X \times_{k,\sigma} \mathbb{C}$ and similarly for Y , $Y^{(m)}$ etc., then by definition the weight filtration W is obtained as in 3.14. from the Leray spectral sequence

$$(3.15.1) \quad E_2^{p,q} = H^p(\sigma X, R^q j_* \mathbb{Q}) \Rightarrow H^{p+q}(\sigma U, \mathbb{Q}) ,$$

and the Hodge filtration F on $H^n(\sigma U, \mathbb{Q}) \otimes \mathbb{C}$ is just the one given by the Hodge filtration F on $H_{DR}^n(\sigma U^{an})$ (analytic de Rham cohomology of the complex analytic space σU^{an} associated to σU) via the isomorphism

$$(3.15.2) \qquad H^n(\sigma U,\mathbb{Q}) \otimes_{\mathbb{Q}} \mathbb{C} \ \xrightarrow{\sim}\ \mathbb{H}^n(\sigma U^{an}, \Omega^{\cdot}_{\sigma U^{an}}) = H^n_{DR}(\sigma U^{an})$$

induced from the quasi-isomorphism $\underline{\mathbb{C}} \to \Omega^{\cdot}_{\sigma U^{an}}$. This Hodge filtration is defined in the same way as the algebraic Hodge filtration F defined above and compatible with it under the GAGA isomorphism

$$(3.15.3) \qquad H^n_{DR}(\sigma U^{an}) \ \xrightarrow{\sim}\ H^n_{DR}(\sigma U/\mathbb{C})$$

with the algebraic de Rham cohomology.

<u>3.16.</u> Combining these isomorphisms with the base change isomorphism

$$(3.16.1) \qquad H^n_{DR}(\sigma U/\mathbb{C}) \ \xrightarrow{\sim}\ H^n_{DR}(U/k) \otimes_{k,\sigma} \mathbb{C} \quad ,$$

we obtain the comparison isomorphism

$$(3.16.2) \qquad I_{\infty,\sigma} = I^n_{\infty,\sigma}(U) \ :\ H^n_{\sigma}(U) \otimes_{\mathbb{Q}} \mathbb{C} \ \xrightarrow{\sim}\ H^n_{DR}(U) \otimes_{k,\sigma} \mathbb{C} \ .$$

As mentioned, $I_{\infty,\sigma}$ respects the Hodge filtration, and it respects the weight filtration by the canonical isomorphism of spectral sequences (compare [D4] (3.1.7), (3.1.8))

$$\begin{array}{ccccc}
H^p(\sigma X, R^q j_*\mathbb{C}) & \Rightarrow & H^{p+q}(\sigma U,\mathbb{C}) & & \\
\downarrow \wr & & & \searrow^{\sim} & \\
\mathbb{H}^p(\sigma X^{an}, gr^W_q \Omega^{\cdot}_{\sigma X^{an}}\langle\sigma Y^{an}\rangle) & \Rightarrow & \mathbb{H}^{p+q}(\sigma X^{an}, \Omega^{\cdot}_{\sigma X^{an}}\langle\sigma Y^{an}\rangle) & \xrightarrow{\sim} & H^{p+q}_{DR}(\sigma U^{an}) \\
\text{GAGA}\downarrow\wr & & \text{GAGA}\downarrow\wr & & \text{GAGA}\downarrow\wr \\
\mathbb{H}^p(\sigma X, gr^W_q \Omega^{\cdot}_{\sigma X}\langle\sigma Y\rangle) & \Rightarrow & \mathbb{H}^{p+q}(\sigma X, \Omega^{\cdot}_{\sigma X}\langle\sigma Y\rangle) & \xrightarrow{\sim} & H^{p+q}_{DR}(\sigma U) \, .
\end{array}$$

The first isomorphism follows from the quasi-isomorphisms

$$(\Omega^{\cdot}_{\sigma X^{an}}\langle\sigma Y^{an}\rangle, W) \ \xleftarrow{id}\ (\Omega^{\cdot}_{\sigma X^{an}}\langle\sigma Y^{an}\rangle, \tau) \hookrightarrow (j_*\Omega^{\cdot}_{\sigma U^{an}}, \tau)$$

of filtered complexes, as the quasi-isomorphism $\mathbb{C} \to \Omega_{\sigma U^{an}}$ induces an isomorphism

$$R^q j_*\mathbb{C} \xrightarrow{\sim} Gr_q^{\tau} j_*\Omega^{\cdot}_{\sigma U^{an}}[q] = \mathcal{H}^n(j_*\Omega^{\cdot}_{\sigma U^{an}}) .$$

3.17. The latter combined with the quasi-isomorphisms

$$Gr_q^W \Omega^{\cdot}_{\sigma X^{an}}\langle\sigma Y^{an}\rangle[q] \to (i_q)_*\Omega^{\cdot}_{(\sigma Y^{(q)})^{an}} \leftarrow (i_q)_*\mathbb{C}$$

gives an isomorphism of analytic sheaves

(3.17.1) $$R^q j_*\mathbb{C} \xrightarrow{\sim} (i_q)_*\mathbb{C}$$

Deligne shows in [D4] (3.1.9) that this induces an isomorphism

(3.17.2) $$\Phi_q^{an} : R^q j_*\mathbb{Q} \xrightarrow{\sim} (i_q)_*\mathbb{Q}(-q)$$

of $\mathbb{Q}$-structures, where $\mathbb{Q}(-q) = \mathbb{Q} \cdot (2\pi i)^{-q} \subseteq \mathbb{C}$. This and the considerations in [D4] (3.2.7) - (3.2.10) show that we have a spectral sequence

(3.17.3) $${}_W E_2^{p,q} = H^p(\sigma Y^{(q)},\mathbb{Q})(-q) \Rightarrow H^{p+q}(\sigma U,\mathbb{Q})$$

of mixed $\mathbb{Q}$-Hodge structures, whose complexification can be identified with the spectral sequence (3.12.3)

$${}_W E_2^{p,q} = \mathbb{H}^p(\sigma Y^{(q)},\Omega^{\cdot}_{\sigma Y^{(q)}})(-q) \Rightarrow \mathbb{H}^{p+q}(\sigma X,\Omega^{\cdot}_{\sigma X}\langle\sigma Y\rangle) = H^{p+q}_{DR}(\sigma U)$$

via the comparison isomorphisms for σU and those of $\sigma Y^{(q)}$ (regarding the $(2\pi i)^{-q}$) .

3.18. $I_{1,\bar\sigma} = I^n_{1,\bar\sigma}(U) : H^n_{\sigma}(U)\otimes\mathbb{Q}_1 \xrightarrow{\sim} H^n_1(U)$ for $\bar\sigma:\bar{k} \hookrightarrow \mathbb{C}$ is defined to be the composition of the canonical comparison isomorphism between complex and étale cohomology

(3.18.1) $$H^n(\sigma U,\mathbb{Q}_1) \xrightarrow{\sim} H^n_{et}(\sigma U,\mathbb{Q}_1)$$

with the isomorphisms

(3.18.2) $$H^n(\sigma U,\mathbb{Q}_l) \cong H^n(\sigma U,\mathbb{Q}) \otimes_{\mathbb{Q}} \mathbb{Q}_l$$

(3.18.3) $$H^n_{et}(U \times_{k,\sigma} \mathbb{C},\mathbb{Q}_l) \overset{\bar{\sigma}^*}{\underset{\sim}{\leftarrow}} H^n_{et}(U \times_k \bar{k},\mathbb{Q}_l) = H^n_l(U) \ .$$

For $\rho \in G_k$ we have $\bar{\sigma}|_k = \sigma = \bar{\sigma}\rho|_k$ and therefore a commutative diagram

$$\begin{array}{ccccccc} I_{l,\bar{\sigma}}: & H^n(\sigma U,\mathbb{Q}_l) & \overset{\sim}{\to} & H^n_{et}(\sigma U,\mathbb{Q}_l) & \overset{\bar{\sigma}^*}{\longleftarrow} & H^n_{et}(U\times_k\bar{k},\mathbb{Q}_l) \\ & \| & & \| & & \uparrow \rho^* \\ I_{l,\bar{\sigma}\rho}: & H^n(\sigma U,\mathbb{Q}_l) & \overset{\sim}{\to} & H^n_{et}(\sigma U,\mathbb{Q}_l) & \overset{(\bar{\sigma}\rho)^*}{\longleftarrow} & H^n_{et}(U\times_k\bar{k},\mathbb{Q}_l) \ . \end{array}$$

$I_{l,\bar{\sigma}}$ respects the weight filtrations as there is an isomorphism of spectral sequences

$$\begin{array}{ccc} E_2^{p,q} = H^p_{et}(\sigma X, R^q j_* \mathbb{Q}_l) & \Rightarrow & H^{p+q}_{et}(\sigma U,\mathbb{Q}_l) \\ \downarrow\wr & & \downarrow\wr \\ E_2^{p,q} = H^p(\sigma X, R^q j_* \mathbb{Q}) \otimes \mathbb{Q}_l & \Rightarrow & H^{p+q}(\sigma U,\mathbb{Q}) \otimes \mathbb{Q}_l \ , \end{array}$$

by the comparison theorem between complex and étale cohomology for constructible sheaves, see [SGA 4] XV 4 . Note that $R^q j_* \mathbb{Q}_l$ is constructible by [SGA 4] XV 5.1 .

<u>3.19. Definition</u> The mixed realization $H^n(U)$ of a smooth quasi-projective variety U over k is

$$H^n(U) = (H^n_{DR}(U), H^n_l(U), H^n(U);\ I_{\infty,\sigma}, I_{l,\bar{\sigma}})_{\substack{l \text{ prime} \\ \sigma: k \hookrightarrow \mathbb{C} \\ \bar{\sigma}: \bar{k} \hookrightarrow \mathbb{C}}}$$

where $H^n_{DR}(U)$ is defined by 3.1, 3.10 and 3.11, $H^n_l(U)$ by 3.14, $H^n_\sigma(U)$ by 3.15 , $I_{\infty,\sigma}$ by 3.16 and $I_{l,\bar{\sigma}}$ by 3.18 . Clearly, this assignment is functorial, so we get the desired functors $(n \in \mathbb{Z})$

$$H^n : \overset{\circ}{\underline{V}}_k \rightarrow \underline{MR}_k$$
$$U \mapsto H^n(U) .$$

In the following we write $\bar{U}$, $\bar{j}$ etc. for the base extensions $U \times_k \bar{k}$, $j \times id_{\bar{k}}$ etc.

3.20. Proposition There are canonical isomorphisms of l-adic sheaves

$$(3.20.1) \qquad \phi_q^{et} : R^q\bar{j}_*\mathbb{Q}_l \ \overset{\sim}{\rightarrow} \ (\bar{i}_q)_*\mathbb{Q}_l(-q)$$

such that via these the spectral sequence (3.14.1) giving the weight filtration on $H_l^n(U)$ can be identified with

$$(3.20.2) \qquad {}_wE_2^{p,q} = H_{et}^p(\overline{Y^{(q)}},\mathbb{Q}_l)(-q) \Rightarrow H_{et}^{p+q}(\bar{U},\mathbb{Q}_l)$$

(note that the $\bar{i}_q$ are acyclic for the étale cohomology), where the differentials

$$d_2^{p,q} : H_{et}^p(\overline{Y^{(q)}},\mathbb{Q}_l)(-q) \rightarrow H_{et}^{p+2}(\overline{Y^{(q-1)}},\mathbb{Q}_l)(-q-1)$$

are given as follows: Let

$$(\bar{\delta}_j)_* : H_{et}^p(\overline{Y^{(q)}},\mathbb{Q}_l)(-q) \rightarrow H_{et}^{p+2}(\overline{Y^{(q-1)}},\mathbb{Q}_l)(-q+1)$$

be the Gysin morphism induced by the closed immersions

$$\delta_j : Y_{i_1 \ldots i_q} \hookrightarrow Y_{i_1 \ldots \hat{i}_j \ldots i_q} .$$

Then $\quad d_2^{p,q} = \sum_{j=1}^{q} (-1)^j(\bar{\delta}_j)_* .$

Proof We closely follow the arguments of Rapoport and Zink in [RZ] §2. Let $\delta_j^r : Y^{(r+1)} \rightarrow Y^{(r)}$, $1 \leq j \leq r+1$, be the map induced by the inclusions

$$Y_{i_1} \cap \ldots \cap Y_{i_{r+1}} \hookrightarrow Y_{i_1} \cap \ldots \cap \hat{Y}_{i_j} \cap \ldots \cap Y_{i_{r+1}}$$

for $1 \leq i_1 < \ldots < i_{r+1} \leq N$. Then we have $a_r \delta_j = a_{r+1}$, where $a_r : Y^{(r)} \to Y$ is the canonical map. By adjunction we get functorial morphisms

$$(\delta_j^r)_*(\delta_j^r)^! F \to F$$

for any étale sheaf on $Y^{(r)}$, and therefore morphisms

$$\partial_j : (a_{r+1})_*(a_{r+1})^! F \to (a_r)_*(a_r)^! F$$

for any sheaf F on Y .

<u>3.20.3. Lemma</u> If I is injective on Y , then

$$\ldots \to (a_{r+1})_*(a_{r+1})^! I \xrightarrow{\partial} (a_r)_*(a_r)^! I \to \ldots \to (a_1)_*(a_1)^! I \to I \to 0$$

is exact, where $\partial = \Sigma(-1)^j \partial_j$.

<u>Proof</u> As in [RZ] Lemma 2.5.

Now let $I^{\cdot}$ be an injective resolution of the constant sheaf $\mathbb{Q}_l$ on $\bar{X}$. There is an exact sequence

(3.20.4) $$0 \to \bar{i}_* \bar{i}^! I^{\cdot} \to I^{\cdot} \to \bar{j}_* \bar{j}^* I^{\cdot} \to 0 .$$

By applying the above lemma to $\bar{i}^! I^{\cdot}$ and by (3.20.4) we get a resolution (note $i_m = i\, a_m$)

(3.20.5) $$\ldots \to (\bar{i}_m)_*(\bar{i}_m)^! I^{\cdot} \to \ldots \to (\bar{i}_1)_*(\bar{i}_1)^! I^{\cdot} \to I^{\cdot} \to$$
$$\to \bar{j}_* \bar{j}^* I^{\cdot} \to 0 .$$

The total complex $sC^{\cdot\cdot}$ associated to the double complex

$$C^{p,q} = \begin{cases} (\bar{i}_{-q})_*(\bar{i}_{-q})^! \; I^p & q \leq -1 \\ I^p & q = 0 , \end{cases}$$

with the differentials induced from (3.20.5), is therefore quasi-isomorphic to $\bar{j}_*\bar{j}^*I^\cdot$.

Now $(\bar{i}_q)^! I^\cdot = \mathbb{R}(\bar{i}_q)^! \mathbb{Q}_1$ is quasi-isomorphic to $\mathbb{Q}_1(-q)[-2q]$ by purity, therefore $C^{\cdot\cdot}$ is quasi-isomorphic to the following double complex $\overline{C^{\cdot\cdot}}$

$$\begin{array}{cccccc}
C^{0,0} & C^{1,0} & C^{2,0} & C^{3,0} & C^{4,0} & C^{5,0} \\
0 & 0 & C^{2,-1}/\mathrm{Im}\,\partial' & C^{3,-1} & C^{4,-1} & C^{5,-1} \\
0 & 0 & 0 & 0 & C^{4,-2}/\mathrm{Im}\,\partial' & C^{5,-2} .
\end{array}$$

As $\overline{C^{p,q}} = 0$ for $p + 2q < 0$, we have

$$\beta_r \, s\,\overline{C^{\cdot\cdot}} := \bigoplus_{q \geq -r} \overline{C^{p,q}} \supseteq \bigoplus_{p+q \leq r} \overline{C^{p,q}} \supseteq \tau_r s\,\overline{C^{\cdot\cdot}} ,$$

τ_r the canonical and β_r the <u>second</u> (increasing) filtration of $s\,\overline{C^{\cdot\cdot}}$.

<u>3.20.6. Lemma</u> $(s\,\overline{C^{\cdot\cdot}}, \tau) \xrightarrow{id} (s\,\overline{C^{\cdot\cdot}}, \beta)$ is a quasi-isomorphism of filtered complexes.

<u>Proof</u> As in [RZ] Lemma 2.7.

This induces quasi-isomorphisms

$$\begin{aligned}
R^q\bar{j}_*\mathbb{Q}_1 &\to (\mathrm{Gr}^q_\tau\, \bar{j}_*\bar{j}^*I^\cdot)[q] \\
&\to (\mathrm{Gr}^q_\beta\, s\,\overline{C^{\cdot\cdot}})[q] = ((\bar{i}_q)_*\bar{i}^!_q I^\cdot[q])[q] \\
&= (\bar{i}_q)_*\mathbb{R}\,\bar{i}^!_q\mathbb{Q}_1[2q] \to (\bar{i}_q)_*\mathbb{Q}_1(-q)
\end{aligned}$$

and therefore the wanted isomorphisms ϕ^{et}_q. Furthermore, the Leray spectral sequence (3.14.1) can - up to renumbering $\tilde{E}^{p,q}_1 = E^{2p+q,-p}_2$ - be identified with the spectral sequence for the second filtration, where the differentials

$$\tilde{d}_1^{p,q} : \mathbb{H}^q(\bar{X}, C^{\cdot,p}) \to \mathbb{H}^q(\bar{X}, C^{\cdot,p+1})$$

are induced by the morphism $C^{\cdot,p} \to C^{\cdot,p+1}$. As $\mathbb{H}^q(\bar{X}, C^{\cdot,p})$ $\cong \mathbb{H}^q(\bar{X}, (\bar{i}_{-p})_* \mathbb{R}(\bar{i}_{-p})^! \mathbb{Q}_1) \cong \mathbb{H}^q(\bar{X}, (\bar{i}_{-p})_* \mathbb{Q}_1(p)[2p])$ $\cong H^{2p+q}(\overline{Y^{(-p)}}, \mathbb{Q}_1)(-p)$, we see that $d_2^{p,q}$ is induced by

$$(\bar{i}_q)_*(\bar{i}_q)^! I^\cdot \xrightarrow{\partial = \Sigma(-1)^j \partial_j} (\bar{i}_{q-1})_*(\bar{i}_{q-1})^! I^\cdot \quad .$$

Via the quasi-isomorphisms $(\bar{i}_r)_*(\bar{i}_r)^! I^\cdot = (\bar{i}_r)_* \mathbb{R}(\bar{i}_r)^! \mathbb{Q}_1$ $\to (\bar{i}_r)_* \mathbb{Q}_1(-r)[-2r]$ this gives the alternating sum of the Gysin morphisms as claimed, as $i_q = \delta_j^{q-1} i_{q-1}$, and the Gysin morphism for $\overline{\delta_j^r} : \overline{Y^{(r+1)}} \to \overline{Y^{(r)}}$ is given by the quasi-isomorphism

$$(\overline{\delta_j^r})_* \mathbb{Q}_1(-1)[-2] \to (\overline{\delta_j^r})_* \mathbb{R}(\overline{\delta_j^r})^! \mathbb{Q}_1$$

followed by the adjunction

$$(\overline{\delta_j^r})_* \mathbb{R}(\overline{\delta_j^r})^! \mathbb{Q}_1 \to \mathbb{Q}_1 \ .$$

In more down to earth terms: if $J^\cdot$ is an injective resolution of $\mathbb{Q}_1$ on $\overline{Y^{(r)}}$ then the Gysin morphism is given by the isomorphism

$$\mathbb{Q}_1(-1) \xrightarrow{\sim} R^2(\overline{\delta_j^r})^! \mathbb{Q}_1 \ ,$$

the quasi-isomorphism

$$R^2(\overline{\delta_j^r})^! \mathbb{Q}_1 = \mathcal{H}^2((\overline{\delta_j^r})^! J^\cdot) \to (\overline{\delta_j^r})^! J^\cdot[2],$$

and the canonical map

$$(\overline{\delta_j^r})_*(\overline{\delta_j^r})^! J^\cdot \to J^\cdot \quad .$$

In cohomology this corresponds to

$$H^p(\overline{Y^{(r)}}, \mathbb{Q}_1(-1)) \cong H^p(\overline{Y^{(r)}}, R^2(\overline{\delta_j^r})^! \mathbb{Q}_1)$$

$$\cong H^{p+2}_{Y^{(r+1)}}(\overline{Y^{(r)}}, \mathbb{Q}_1) \xrightarrow{can} H^{p+2}(\overline{Y^{(r)}}, \mathbb{Q}_1) \ ,$$

(the second isomorphism from the spectral sequence for

$(\overline{\delta_j^r})^!)$, at least for $r \neq 0$. For $r = 0$, i.e., $Y^{(0)} = X$ the map $\delta^{(0)}$ is not a closed immersion, and one has to restrict to the $Y_i \subseteq Y^{(1)}$.

<u>3.21. Lemma</u> a) The differentials $d_2^{p,q}$ in the spectral sequences (3.12.3) and (3.17.3) are given as alternating sums of Gysin morphisms as in proposition 3.20.

b) The isomorphism (3.17.2) of analytic sheaves

$$\phi_q^{an} : R^q(\sigma j)_*\mathbb{Q} \xrightarrow{\sim} (\sigma i_q)_*\mathbb{Q}(-q)$$

corresponds to ϕ_q^{et} via the comparison isomorphism for constructible (smooth) sheaves for the étale and the complex analytic topology [SGA 4] XVI 4.1.

c) There are canonical isomorphisms of spectral sequences

$$\begin{array}{ccccc}
E_2^{p,q} & = & H_{et}^p(\overline{Y^{(q)}},\mathbb{Q}_l)(-q) & \Rightarrow & H_{et}^{p+q}(\bar{U},\mathbb{Q}_l) \\
 & & \downarrow \wr\, \bar{\sigma}^* & & \downarrow \wr\, \bar{\sigma}^* \\
E_2^{p,q} & = & H_{et}^p(\sigma Y^{(q)},\mathbb{Q}_l)(-q) & \Rightarrow & H_{et}^{p+q}(\sigma U,\mathbb{Q}_l) \\
 & & \uparrow \wr & & \uparrow \wr \\
E_2^{p,1} & = & H^p(\sigma Y^{(q)},\mathbb{Q}(-q)) \otimes \mathbb{Q}_l & \Rightarrow & H^{p+q}(\sigma U,\mathbb{Q}) \otimes \mathbb{Q}_l
\end{array}$$

given by the comparison isomorphism between complex and étale cohomology, which correspond to the isomorphisms of the Leray spectral sequences via the isomorphisms ϕ_q^{et} and ϕ_q^{an}.

<u>Proof</u> a) seems to be well-known, though I could not find a good reference. The claim for the de Rham cohomology can be checked via the definition of the isomorphism (3.6.1) and the explicit formula for the Gysin morphism given in [Ber] VI 3.1.3. This implies the claim for the singular cohomology, as Gysin morphisms

correspond under the comparison isomorphisms.

Another approach and probably the most natural proof of b) is given by the observation that the whole construction in the proof of lemma 3.20 can be carried out for analytic sheaves on σX^{an}. Then we get isomorphisms

$$'\phi_q^{an} : R^q(\sigma j)_*\mathbb{Q} \xrightarrow{\sim} (\sigma i_q)_*\mathbb{Q}(-q)$$

and a canonical spectral sequence

$$(3.21.1) \qquad E_2^{p,q} = H^p(\sigma Y^{(q)},\mathbb{Q})(-q) \Rightarrow H^{p+q}(\sigma U,\mathbb{Q})$$

which is isomorphic to the Leray spectral sequence for σj, and where the differentials $d_2^{p,q}$ are alternating sums of Gysin morphisms as in lemma 3.20. Moreover, after tensoring with $\mathbb{Q}_1$ and applying the functor ε^* associating to each étale sheaf the corresponding complex analytic sheaf (see [SGA 4] XVI 4.1; strictly speaking, we have to consider $\mathbb{Z}/1^n\mathbb{Z}$-sheaves and then pass to limits and $\mathbb{Q}_1$-"sheaves"), we can compare the whole process on each step with the étale construction via the canonical base change morphisms. As these give isomorphisms for constructible sheaves and as ε^* is exact, we see that $'\phi_q^{an}$ and ϕ_q^{et} are compatible under the comparison isomorphism, i.e.,

$$\begin{array}{ccc}
\varepsilon^* R^q(\sigma j^{et})_*\mathbb{Q}_1 & \xrightarrow[\sim]{\varepsilon^*\phi_q^{et}} & \varepsilon^*(\sigma i_q^{et})_*\mathbb{Q}_1(-q) \\
\text{base change} \downarrow \wr & & \wr \downarrow \text{base change} \\
R^q(\sigma j^{an})_*\mathbb{Q}_1 & \xrightarrow[\sim]{'\phi_q^{an}\otimes\mathbb{Q}_1} & (\sigma i_q^{an})_*\mathbb{Q}_1(-q)
\end{array}$$

is commutative, and there is an isomorphism between the spectral sequences (3.20.2) and (3.21.1) $\otimes\ \mathbb{Q}_1$ via the comparison (base change) isomorphisms.

It remains to prove $'\phi_q^{an} = \phi_q^{an}$. There is an isomorphism

$$(\sigma i_q^{an})_* \mathbb{Q} \xrightarrow{\sim} \overset{q}{\wedge} (\sigma i_1^{an})_* \mathbb{Q}$$

depending on the fixed ordering of the Y_j , and a canonical isomorphism

$$R^q(\sigma j^{an})_* \mathbb{Q} \underset{\sim}{\rightarrow} \overset{q}{\wedge} R^1(\sigma j^{an})_* \mathbb{Q}$$

given by the cup-product (compare [D4] 3.1). Both ϕ_q^{an} and $'\phi_q^{an}$ are compatible with these isomorphisms, so we only have to show $'\phi_1^{an} = \phi_1^{an}$. The question is local, and we can replace σX by an open polycylinder D^d , with $D = \{z \in \mathbb{C} | \; |z| < 1\}$, and suppose that $\sigma Y = \bigcup_{j=1}^{\nu} \sigma Y_j$ with $\sigma Y_j = pr_j^{-1}(0)$ such that $\sigma U = (D^\times)^\nu \times D^{d-\nu}$, with $D^\times = D \setminus \{0\}$. Then the fibre at 0 of $R^q(\sigma j^{an})_* \mathbb{Q}$ is isomorphic to $H^q(\sigma U, \mathbb{Q})$, and by definition $(\phi_1^{an})^{-1}$ maps the class of σY_j in the fibre of $(\sigma i_1^{an})_* \mathbb{Q}$ at 0 to the class of $\frac{dx_j}{x_j}$ in $H^1(\sigma U, \mathbb{Q})(1)$, where x_j are the canonical coordinates in D^d . In the same setting, the canonical generator of

$$H^2_{\sigma Y_j}(\sigma X, \mathbb{Q})(1) \cong \mathbb{Q}$$

is the image under the connecting morphism

$$H^1(\sigma X \setminus \sigma Y_j, \mathbb{Q})(1) \xrightarrow{\sim} H^2_{\sigma Y_j}(\sigma X, \mathbb{Q})(1)$$

of the canonical generator α_j of

$$H^1(\sigma X \setminus \sigma Y_j, \mathbb{Q} \cdot 2\pi i) = \mathrm{Hom}(\pi_1(\sigma X \setminus \sigma Y_j), \mathbb{Q} \cdot 2\pi i)$$

which sends the generating path γ_j around σY_j which has positive orientation (w.r.t. the orientation of σX given by the choice of $i = \sqrt{-1}$) to $2\pi i$. Now $'\phi_1^{an}$ sends the class of σY_j to the image of α_j under the restriction map

$$H^1(\sigma X \setminus \sigma Y_j, \mathbb{Q}(1)) \rightarrow H^1(\sigma U, \mathbb{Q}(1)) ,$$

and as

$$\int_{\gamma_j} \frac{dx_j}{x_j} = 2\pi i$$

we see that this image coincides with the class of $\frac{dx_j}{x_j}$, i.e., we have shown $'\phi_1^{an} = \phi_1^{an}$.

3.22. From the above it is clear that we have a spectral sequence of realizations

$$(3.22.1) \qquad {}_W E_2^{p,q} = H^p(Y^{(q)})(-q) \Rightarrow H^{p+q}(U)$$

giving the weight filtration on the realization attached to U , where the differentials $d_2^{p,q}$ are given by alternating sums of Gysin morphisms. In particular, $Gr_{n+k}^W\ H^n(U)$ is isomorphic to a subquotient of $H^{n-k}(Y^{(k)})(-k)$, namely to $\mathrm{Ker}\ d_2^{n-k,k}/\mathrm{Im}\ d_2^{n-k-2,k+1}$, as the spectral sequence (3.22.1) degenerates at the E_3-terms. The latter has only to be proved for one realization, and is proved for the Hodge realization by Deligne [D4] (3.2.13).

§4. The category of mixed motives

With the notations of the previous section we define the functor

$$H: \ \underline{\overset{\circ}{V}}_k \rightarrow \underline{MR}_k$$

by $H(U) = \bigoplus_{n \geq 0} H^n(U)$ for U smooth quasi-projective over k .

4.1. Definition The category $\underline{MM}_k$ of mixed motives (for absolute Hodge cycles) over k is the Tannakian subcategory of $\underline{MR}_k$ generated by the image of H .

4.2. Proposition A mixed realization $H \in \underline{MR}_k$ is a mixed motive if and only if it is a subquotient of $H(U) \otimes H(V)^{\vee} =$

$\underline{\mathrm{Hom}}(H(V),H(U))$ for some smooth quasi-projective varieties U and V over k.

Proof a) We first show that $\underline{MM}_k$ contains all these sub-quotients. Let $U,V \in \overset{o}{\underline{V}}_k$ and $n \in \mathbb{Z}$.

i) $H^n(U) \in \underline{MM}_k$ as a direct factor of $H(U)$, i.e., as kernel of an idempotent in $\mathrm{End}(H(U))$.

ii) $W_mH^n(U) \in \underline{MM}_k$ for all $m \in \mathbb{Z}$ by induction on m: for $m >> 0$ $W_mH^n(U) = H^n(U)$, and $W_mH^n(U) \in \underline{MM}_k$ implies $W_{m-1}H^n(U) = \mathrm{Ker}(W_mH^n(U) \to Gr_m^WH^n(U)) \in \underline{MM}_k$, as $Gr_m^WH^n(U) \in \underline{MM}_k$ for all m by 3.22 and lemma 1.1: By 3.22 $Gr_m^WH^n(U)$ is isomorphic to a sub-quotient of $H^{n'}(Y)(n")$ for some smooth projective variety Y over k and some $n',n" \in \mathbb{Z}$, and we can reformulate lemma 1.1 in the language of realizations:

4.3. Lemma If Y is smooth projective over k, then any subquotient of $H^n(Y)(r)$, $n,r \in \mathbb{Z}$, is a direct factor.

So $Gr_m^WH^n(U)$ is a direct factor of $H^{n'}(Y)(n")$, which is a mixed motive by the following remark.

iii) If $H \in \underline{MM}_k$, then $H(r) \in \underline{MM}_k$ for all $r \in \mathbb{Z}$.
In fact, $1(-1)$ is canonically isomorphic to $H^2(\mathbb{P}^1)$, and $\underline{MM}_k$ is closed under formation of tensor products and duals.

iv) By the last argument also $W_m(H^n(V)^\vee) = (W_{-m}H^n(V))^\vee \in \underline{MM}_k$ and so $W_m(H(U) \otimes H(V)^\vee) \in \underline{MM}_k$ for all $m \in \mathbb{Z}$.

v) If $H \subseteq H_o = H(U) \otimes H(V)^\vee$ we show $H \in \underline{MM}_k$ by induction over the number of $m \in \mathbb{Z}$ with $Gr_m^WH \neq 0$. Let $m = \max\{n\in\mathbb{Z} \mid Gr_n^WH \neq 0\}$ and consider the cartesian square

$$\begin{array}{ccc} H' & \twoheadrightarrow & Gr_m^WH \\ \cap| & & \cap| \\ W_mH_o & \twoheadrightarrow & Gr_m^WH_o \end{array} .$$

Then the quotient $Gr_m^W H_o / Gr_m^W H$ is in $\underline{MM}_k$, as it is a direct factor of $Gr_m^W H_o \in \underline{MM}_k$ by lemma 1.1 or rather 4.3. Namely $Gr_m^W H_o$ is a subquotient of $H^{n'}(Y)(n")$ for some smooth projective Y and $n',n" \in \mathbb{Z}$ by the same arguments as in ii), as $H^n(Y_2)^\vee \cong H^{2\dim Y_2 - n}(Y_2)(\dim Y_2)$ by Poincaré duality, $H^{n_1}(Y_1) \otimes H^{n_2}(Y_2) \subseteq H^{n_1+n_2}(Y_1 \times Y_2)$ by the Künneth formula, $H^n(Y_1)(-1) \subseteq H^{n+2}(Y_1 \times \mathbb{P}^1)$ and $H^n(Y_1) \oplus H^n(Y_2) \cong H^n(Y_1 \amalg Y_2)$ for smooth projective varieties Y_i and $n_i \in \mathbb{Z}$.

Therefore $H' \in \underline{MM}_k$, as it is the kernel of the map $W_m H_o \to Gr_m^W H_o / Gr_m^W H$. By induction $W_{m-1}H$ is in $\underline{MM}_k$, and from the commutative exact diagram

$$\begin{array}{ccccccccc}
 & & \bar{H} & = & \bar{H} & & & & \\
 & & \uparrow & & \uparrow & & & & \\
0 & \longrightarrow & W_{m-1}H_o & \dashrightarrow & H' & \longrightarrow & Gr_m^W H & \longrightarrow & 0 \\
 & & \uparrow & & \uparrow & & \| & & \\
0 & \longrightarrow & W_{m-1}H & \longrightarrow & H & \longrightarrow & Gr_m^W H & \longrightarrow & 0
\end{array}$$

we see that $H = \mathrm{Ker}(H' \to \bar{H} = W_{m-1}H_o / W_{m-1}H)$ is in $\underline{MM}_k$.

b) Now we have to show that the full subcategory consisting of all subquotients of $H(U) \otimes H(V)^\vee$ for U and $V \in \overset{o}{\underline{V}}_k$ is a Tannakian subcategory of $\underline{MR}_k$.

i) Let B/A be a subquotient of $H \in \underline{MR}_k$, $A \subseteq B \subseteq H$, and B'/A' a subquotient of $H' \in \underline{MR}_k$, $A' \subseteq B' \subseteq H'$. Then $B/A \otimes B'/A'$ is (isomorphic to) a subquotient of $H \otimes H'$, via the canonical isomorphism

$$B \otimes B' / A \otimes B' + A' \otimes B \quad \overset{\sim}{\to} \quad B/A \otimes B'/A' \quad .$$

ii) $(H(U_1) \otimes H(V_1)^\vee) \otimes (H(U_2) \otimes H(V_2)^\vee)$ is isomorphic to $H(U_1 \times U_2) \otimes H(V_1 \times V_2)^\vee = H(U_1) \otimes H(U_2) \otimes H(V_1)^\vee \otimes H(V_2)^\vee$.

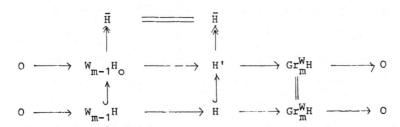

iii) If B/A is a subquotient of $H \in \underline{MR}_k$, then $(B/A)^{\vee} = A^{\vee}/B^{\vee}$ is a subquotient of $H^{\vee}$, and $(H(U) \otimes H(V)^{\vee})^{\vee} \cong H(V) \otimes H(U)^{\vee}$.

Now the full subcategory of $\underline{MR}_k$ formed by the above subquotients is obviously abelian, and the above remarks show that it is closed under formation of tensor products and duals. Finally it contains the identity object $1 \in \underline{MR}_k$, which can be identified with $H(\mathrm{Spec}\ k)$, so it is indeed a Tannakian subcategory of $\underline{MR}_k$. q.e.d.

4.4. Theorem a) The functor $H: \underline{M}_k \to \underline{R}_k$ which to any motive (for absolute Hodge cycles) over k associates its realization, is a fully faithful tensor functor and identifies $\underline{M}_k$ with the full subcategory $\underline{M}'_k$ of $\underline{MM}_k$, whose objects are direct sums of pure realizations, i.e., with the categorial intersection $\underline{MM}_k \cap \underline{R}_k$ in $\underline{MR}_k$.
b) If $\underline{V}_k$ and $\mathring{\underline{V}}_k$ are the categories of smooth projective and smooth quasi-projective varieties, respectively, we have a commutative diagram of functors

$$\begin{array}{ccccc}
 & & \underline{V}_k & \hookrightarrow & \mathring{\underline{V}}_k \\
 & \swarrow^{h} & \downarrow H|_{\underline{V}_k} & & \downarrow H \\
\underline{M}_k & \xrightarrow[\sim]{H} & \underline{M}'_k & \hookrightarrow & \underline{MM}_k \\
 & & \downarrow & & \downarrow \\
 & & \underline{R}_k & \hookrightarrow & \underline{MR}_k
\end{array}$$

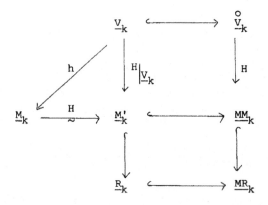

where $\hookrightarrow$ means a fully faithful functor giving an embedding of a subcategory which is closed under formation of subquotients, and where $\underset{\sim}{\rightarrow}$ means an equivalence of categories.

c) If we identify $\underline{M}_k$ with its image M_k' under H , which we will always do from now on, then the category $\underline{M}_k$ of motives is also the Tannakian subcategory of $\underline{MR}_k$ which is generated by the image of

$$H_{|\underline{V}_k} : \underline{V}_k \rightarrow \underline{MR}_k \quad .$$

d) $H \in \underline{MR}_k$ is a motive if and only if it is a subquotient of H(X)(r) for some smooth projective variety X over k and some $r \in \mathbb{Z}$. It is then also a direct summand of H(X)(r) .

<u>Proof</u> a) and d): H: $\underline{M}_k \rightarrow \underline{MR}_k$ is the unique functor making the left triangle in the diagram of b) commutative, and it was already mentioned that it is fully faithful (compare [DMOS] II 6.7 g)). The constraints of the tensor category $\underline{M}_k$ are just choosen in such a way (passing from "false motives" to "true motives", see [DMOS] II p. 203) that H becomes a tensor functor.

For a motive M over k , the realization H(M) is a direct factor of H(X)(r) for some smooth projective X/k and some $r \in \mathbb{Z}$, compare [DMOS] II 6.7 b) . In particular, H(M) is a sum of pure realizations, i.e., lies in $\underline{R}_k$, and it is of the form stated in d) . Furthermore we have

$$H(X)(r) \cong \begin{cases} H(X) \otimes H^2(\mathbb{P}^1)^{\otimes|r|} & r \leq 0 \\ H(X) \otimes (H^2(\mathbb{P}^1)^{\vee})^{\otimes r} & r > 0 \, , \end{cases}$$

so H(X)(r) and therefore also H(M) lies in $\underline{MM}_k$. Conversely, any subquotient of H(X)(r) "is" a motive by lemma 1.1/4.3, which shows d.).

Finally, let H be an object of $\underline{MM}_k \cap \underline{R}_k$, given as a subquotient B/A for $A \subseteq B \subseteq H(U) \otimes H(V)^{\vee}$, where U, V are smooth quasi-projective over k . By assumption, $B/A \cong \bigoplus_{m \in \mathbb{Z}} Gr_m^W(B/A)$, so we only have to show $Gr_m^W(B/A) \in H(\underline{M}_k)$ for all $m \in \mathbb{Z}$. But $Gr_m^W(B/A)$ is a subquotient of $Gr_m^W(H(U) \otimes H(V)^{\vee}) = \bigoplus_{p+q=m} Gr_p^W H(U) \otimes Gr_q^W H(V)^{\vee}$ so by using 4.3 as before, we only have to show that $Gr_m^W(H(U) \otimes H(V)^{\vee}) \in H(\underline{M}_k)$ for all $m \in \mathbb{Z}$, i.e., that $Gr_p^W H(U) \in \underline{M}_k$ for all smooth quasi-projective U/k as $\underline{M}_k$ is closed under formation of tensor products and duals, and $Gr_q^W(H(V)^{\vee}) = (Gr_{-q}^W(H(V))^{\vee}$. Choosing $X \supseteq U$ smooth projective and $Y = \bigcup_{i=1}^{N} Y_i$ as in section 3 , we obtain that $Gr_p^W H(U)$ is a direct sum of the $Gr_p^W H^n(U)$, and by 3.22 that $Gr_p^W H^n(U)$ is a subquotient of $H^{2n-p}(Y^{(p-n)})(-p+n)$ and therefore a motive by lemma 4.3.

b) and c): Let $\underline{M}_k''$ be the Tannakian subcategory generated by the image of $H: \underline{V}_k \to \underline{MR}_k$. Then $\underline{M}_k''$ contains every direct summand of $H(X)(r)$ for $X \in \underline{V}_k$ and $r \in \mathbb{Z}$, and so it contains $\underline{M}_k'$. The other inclusion follows from the fact that $\underline{M}_k'$ is a Tannakian subcategory. $\underline{R}_k \hookrightarrow \underline{MR}_k$ is closed w.r.t. subquotients by 2.15, the corresponding statement for $\underline{M}_k \hookrightarrow \underline{MR}_k$ is equivalent to lemma 1.1/4.3, for $\underline{MM}_k \hookrightarrow MR_k$ it follows from 4.2 and the rest is clear.

<u>4.5. Remark</u> In the proof we have seen, that for any mixed motive N the subobjects $W_m N$ are mixed motives and the subquotients $Gr_m^W N$ are motives. In particular, any mixed motive is a successive extension of motives, and the spectral sequence (3.22.1)

$$ {}_W E_2^{p,q} = H^p(Y^{(q)})(-q) \Rightarrow H^{p+q}(U) $$

is a spectral sequence of mixed motives.

4.6 By a basic theorem on Tannakian categories, the neutral Tannakian categories $\underline{M}_k$ and $\underline{MM}_k$ are equivalent to the categories of representations of certain pro-algebraic groups over $\mathbb{Q}$. These arise as automorphism groups of the fibre functors giving a neutralization: For an embedding $\sigma: k \hookrightarrow \mathbb{C}$ let $MG(\sigma)$ be the automorphism group of the fibre functor

$$H_\sigma : \underline{MM}_k \to \underline{Vec}_{\mathbb{Q}}$$

given by the restriction of $H_\sigma : \underline{MR}_k \to \underline{Vec}_{\mathbb{Q}}$ (see 2.13), and $G(\sigma)$ the automorphism group of the restriction

$$H_\sigma : \underline{M}_k \to \underline{Vec}_{\mathbb{Q}} .$$

If $\underline{Rep}\ G$ denotes the category of (finite dimensional) algebraic representations of a pro-algebraic group $G/\mathbb{Q}$, we have equivalences of tensor categories

$$\underline{M}_k \xrightarrow{\sim} \underline{Rep}\ G(\sigma)$$
$$\underline{MM}_k \xrightarrow{\sim} \underline{Rep}\ MG(\sigma) ,$$

and the inclusion $\underline{M}_k \hookrightarrow \underline{MM}_k$ corresponds to a morphism of pro-algebraic groups $\psi: MG(\sigma) \to G(\sigma)$, see [SR]. The inclusion of the category $\underline{M}_k^o$ of Artin motives (see [DMOS] II 6.17) in $\underline{M}_k$ and $\underline{MM}_k$ gives morphisms $\pi: G(\sigma) \to G_k$ and $M\pi : MG(\sigma) \to G_k$, as $\underline{M}_k^o \xrightarrow{\sim} \text{Rep}\ G_k$ (loc. cit.). Finally, the base extension functor $\underline{MR}_k \to \underline{MR}_{\bar{k}}$ (see 2.16 i)) induces base extension functors $\underline{M}_k \to \underline{M}_{\bar{k}}$ and $\underline{MM}_k \to \underline{MM}_{\bar{k}}$, as canonically $H(U) \times_k k' \cong H(U \times_k k')$ for a smooth quasi-projective variety U over k and a field extension k'/k. Thus we get homomorphisms $i : G^o(\sigma) \to G(\sigma)$ and $Mi : MG^o(\sigma) \to MG(\sigma)$, where $G^o(\sigma) = G(\bar{\sigma}) = \underline{Aut}^{\otimes}(H_{\bar{\sigma}}|_{\underline{M}_{\bar{k}}})$ and $MG^o(\sigma) = MG(\bar{\sigma}) = \underline{Aut}^{\otimes}(H_{\bar{\sigma}}|_{\underline{MM}_{\bar{k}}})$ are the automorphism groups of the fibre functor $H_{\bar{\sigma}}$ on $\underline{M}_{\bar{k}}$ and $\underline{MM}_{\bar{k}}$ respectively, for some embedding $\bar{\sigma} : \bar{k} \hookrightarrow \mathbb{C}$ with $\bar{\sigma}|_k = \sigma$.

<u>4.7. Theorem</u> a) ψ is an epimorphism, i.e., faithfully flat.

b) Via ψ, $G(\sigma)$ is identified with the maximal pro-reductive quotient of $MG(\sigma)$; the kernel $U(\sigma)$ of ψ is connected and pro-unipotent.

c) With the above notations (involving a choice of $\bar{\sigma} : \bar{k} \hookrightarrow \mathbb{C}$ with $\bar{\sigma}|_k = \sigma$) there is a commutative exact diagram of pro-algebraic groups

$$\begin{array}{ccccccccc}
 & & U(\sigma) & = & U(\sigma) & & & & \\
 & & \downarrow & & \downarrow & & & & \\
1 & \longrightarrow & MG^{o}(\sigma) & \xrightarrow{Mi} & MG(\sigma) & \xrightarrow{M\pi} & G_k & \longrightarrow & 1 \\
 & & \downarrow & & \psi \downarrow & & \| & & \\
1 & \longrightarrow & G^{o}(\sigma) & \xrightarrow{i} & G(\sigma) & \xrightarrow{\pi} & G_k & \longrightarrow & 1 .
\end{array}$$

d) $G^{o}(\sigma)$ and $MG^{o}(\sigma)$ are connected and the identity components of $G(\sigma)$ and $MG(\sigma)$, respectively. $G^{o}(\sigma)$ is the maximal pro-reductive quotient of $MG^{o}(\sigma)$.

e) For any $\tau \in G_k$, $(M\pi)^{-1}(\tau) = \underline{\mathrm{Hom}}^{\otimes}(H_{\bar{\sigma}}, H_{\bar{\sigma}\tau})$, regarding $H_{\bar{\sigma}}$ and $H_{\bar{\sigma}\tau}$ as functors on $\underline{MM}_k$, and $\pi^{-1}(\tau) = \underline{\mathrm{Hom}}^{\otimes}(H_{\bar{\sigma}}, H_{\bar{\sigma}\tau})$ for the restrictions to $\underline{M}_k$.

f) For any prime l , there are canonical continuous homomorphisms $sp_l : G_k \to G(\sigma)(\mathbb{Q}_l)$ and $Msp_l : G_k \to MG(\sigma)(\mathbb{Q}_l)$ with $\pi \circ sp_l = id$, $M\pi \circ Msp_l = id$ and $sp_l = \psi \circ Msp_l$.

g) There is a canonical section $\Sigma : G(\sigma) \to MG(\sigma)$ of ψ , corresponding to the semi-simplification functor

$$\begin{aligned} s.s. \; : \; \underline{MM}_k &\to \underline{M}_k \\ N &\mapsto \bigoplus_{m \in \mathbb{Z}} Gr^W_m N . \end{aligned}$$

One has $M\pi \circ \Sigma = \pi$, and $\Sigma \circ sp_l = Msp_l$ for any l .

Pictorially:

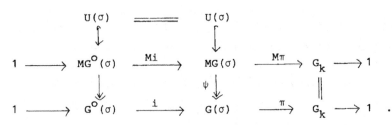

(4.7.1)
$$\begin{array}{ccc} MG(\sigma) & \overset{M\pi}{\underset{Msp_1}{\rightleftarrows}} & G_k \\ \psi\Big\downarrow\Big\uparrow\Sigma & & \Big\| \\ G(\sigma) & \overset{\pi}{\underset{sp_1}{\rightleftarrows}} & G_k \end{array}$$

is commutative in all ways with equi-directed horizontal maps.

<u>Proof</u> a) ψ is faithfully flat, as $\underline{M}_k \hookrightarrow \underline{MM}_k$ is fully faithful and saturated w.r.t. subquotients by 4.4 b), see [DMOS] II 2.21 a).

b) $G(\sigma)$ is pro-reductive, as $\underline{M}_k$ is semi-simple, see [DMOS] II 6.22 (here we use "reductive" also for non-connected groups). Let $\overline{G(\sigma)}$ be the maximal pro-reductive quotient of $MG(\sigma)$. Then $\underline{\mathrm{Rep}}\ \overline{G(\sigma)} \subseteq \underline{\mathrm{Rep}}\ MG(\sigma) = \underline{MM}_k$ is semi-simple and closed with respect to subquotients. For $N \in \underline{\mathrm{Rep}}\ \overline{G(\sigma)}$ we therefore have $N \cong \bigoplus_{m\in\mathbb{Z}} Gr_m^W N$, i.e., $N \in \underline{R}_k$ and therefore $N \in \underline{MM}_k \cap \underline{R}_k = \underline{M}_k = \underline{\mathrm{Rep}}\ G(\sigma)$. We conclude $\underline{\mathrm{Rep}}\ \overline{G(\sigma)} = \underline{\mathrm{Rep}}\ G(\sigma)$ and thus $\overline{G(\sigma)} = G(\sigma)$. With the same arguments, $G^o(\sigma)$ is the maximal pro-reductive quotient of $MG^o(\sigma)$, and by c), $U(\sigma)$ is the pro-unipotent radical of the connected pro-algebraic group $MG^o(\sigma)$.

c) - f) · The statements for $G(\sigma)$ are proved in [DMOS] II 6.23, and the proofs for $MG(\sigma)$ are similar:

c) Mi is a closed immersion as any object N of $\underline{MM}_{\bar{k}}$ is a subquotient of an object $N_o \times_k \bar{k}$ with $N_o \in \underline{MM}_k$ (see the criterion in [DMOS] II 2.21 b)). In fact, it suffices to consider the case $N = H(U) \otimes H(V)^{\vee}$ for $U, V \in \underline{\overset{o}{V}}_{\bar{k}}$; but U and V have models U' and V' over a finite extensions k' of k, and we can take $N_o = R_{k'/k}(H(U') \otimes H(V')^{\vee}) = H(Res_{k'/k}U') \otimes$

$H(\mathrm{Res}_{k'/k}V')^\vee$, where $\mathrm{Res}_{k'/k}U'$ is the Grothendieck restriction $U' \to \mathrm{Spec}\ k' \to \mathrm{Spec}\ k$, as $N_o \times_k k' \overset{p}{\to} H(U') \otimes H(V')^\vee$ is surjective.

$M\pi$ is faithfully flat as $\underline{M}_k^o \hookrightarrow \underline{MM}_k$ is fully faithful and saturated w.r.t. subquotients, and the exactness of $1 \to MG^o(\sigma) \overset{Mi}{\to} MG(\sigma) \overset{M\pi}{\to} G_k \to 1$ is a special case of e) . Finally, $U(\sigma)$ lies in $MG^o(\sigma) = \ker M\pi$, as $M\pi = \pi \circ \psi$ which is clear from the factorization $\underline{M}_k^o \hookrightarrow \underline{M}_k \hookrightarrow \underline{MM}_k$.

d) For the connectedness of $MG^o(\sigma)$ we have to show that for any non-trivial representation X of $MG^o(\sigma)$ the category of subquotients of X^n , $n \geq 0$, is not stable under tensor products, see [DMOS] II 2.22. But such an object $N \in \underline{\mathrm{Rep}}\ MG^o(\sigma) = \underline{MM}_{\bar{k}}$ must be pure of weight zero, as the weights occurring in N^n are the same as those occurring in N, and therefore bounded. In particular $N \in \underline{M}_{\bar{k}} = \underline{\mathrm{Rep}}\ G^o(\sigma)$, and we have reduced to the connectedness of $G^o(\sigma)$, which is proved in [DMOS] II 6.22. As G_k is totally disconnected, $MG^o(\sigma)$ is the full identity component of $MG(\sigma)$.

e) We have to associate to any $g \in MG(\sigma)(R) = \mathrm{Hom}^\otimes(H_\sigma \otimes R, H_\sigma \otimes R)$, R a $\mathbb{Q}$-algebra, a canonical element of $\underline{\mathrm{Hom}}^\otimes(H_{\bar{\sigma}}, H_{\bar{\sigma}\tau})(R) = \mathrm{Hom}^\otimes(H_{\bar{\sigma}} \otimes R, H_{\bar{\sigma}\tau} \otimes R)$ for $\tau = M\pi(g)$. We write $H_\sigma(M,R) = H_\sigma(M) \otimes R$ and $\bar{M} = M \times_k \bar{k}$. Then for $M,N \in \underline{MM}_k$ and $f \in \mathrm{Hom}(\bar{M},\bar{N})$ there is a commutative diagram

$$\begin{array}{ccc} H_{\bar{\sigma}}(\bar{M},R) = H_\sigma(M,R) & \xrightarrow{g_M} & H_\sigma(M,R) = H_{\bar{\sigma}\tau}(\bar{M},R) \\ f_{\bar{\sigma}} \downarrow & & \downarrow f_{\bar{\sigma}\tau} \\ H_{\bar{\sigma}}(\bar{N},R) = H_\sigma(N,R) & \xrightarrow{g_N} & H_\sigma(N,R) = H_{\bar{\sigma}\tau}(\bar{N},R) . \end{array} \qquad (4.7.2)$$

In fact, by applying the functoriality of the g_M to the evaluation map $M \times \underline{\mathrm{Hom}}(M,N) \to N$ one sees that there is a commutative diagram

$$
(4.7.3)\qquad
\begin{array}{ccc}
H_\sigma(M,R) & \xrightarrow{\;g_M\;} & H_\sigma(M,R) \\
{\scriptstyle \tilde{f}_{\bar\sigma}}\downarrow & & \downarrow{\scriptstyle g\tilde{f}_\sigma} \\
H_\sigma(N,R) & \xrightarrow{\;g_N\;} & H_\sigma(N,R)
\end{array}
$$

for any $\tilde{f}_\sigma \in H_\sigma(\underline{\mathrm{Hom}}(M,N)) = \mathrm{Hom}(H_\sigma(M),H_\sigma(N))$, where $g = g_{\underline{\mathrm{Hom}}(M,N)}$. On the other hand, via the action of G_k (see 2.19) $\mathrm{Hom}(\bar{M},\bar{N})$ can be regarded as an Artin motive, i.e., an object of $\underline{M}_k^o$ = $\underline{\mathrm{Rep}}\ G_k$ (compare [DMOS] II 6.17 and 6.18) . This depends on the choice of an extension $\bar\rho : \bar{k} \hookrightarrow \mathbb{C}$ for any $\rho : k \hookrightarrow \mathbb{C}$, and we choose the extension $\bar\sigma$ for σ . Then there is a morphism of mixed motives $\mathrm{Hom}(\bar{M},\bar{N}) \overset{j}{\hookrightarrow} \underline{\mathrm{Hom}}(M,N)$ such that j on

$$H_{DR}(\mathrm{Hom}(\bar{M},\bar{N})) = (\mathrm{Hom}(\bar{M},\bar{N}) \otimes \bar{k})^{G_k}$$

and $$H_1(\mathrm{Hom}(\bar{M},\bar{N}) = \mathrm{Hom}(\bar{M},\bar{N}) \otimes_{\mathbb{Q}} \mathbb{Q}_1$$

is induced by the projection to $\mathrm{Hom}_{\bar{k}}(H_{DR}(\bar{M})\ ,\ H_{DR}(\bar{N}))$ and $\mathrm{Hom}_{\mathbb{Q}_1}(H_1(\bar{M}),H_1(\bar{N}))$, respectively, and on

$$H_\sigma(\mathrm{Hom}(\bar{M},\bar{N})) = \mathrm{Hom}(\bar{M},\bar{N})$$

is the projection to $\mathrm{Hom}(H_{\bar\sigma}(\bar{M}),H_{\bar\sigma}(\bar{N})) = \mathrm{Hom}(H_\sigma(M),H_\sigma(N))$. If $M\pi(g) = \tau$, then by definiton g acts like τ on $\mathrm{Hom}(\bar{M},\bar{N}) \subseteq \underline{\mathrm{Hom}}(M,N)$. So (4.7.2) follows from (4.7.3) , as

$$
\begin{array}{ccccc}
H_{\bar\sigma}(\bar{M}) & = & H_\sigma(M) & = & H_{\bar\sigma\tau}(\bar{M}) \\
{\scriptstyle (\tau f)_{\bar\sigma}}\downarrow & & & & \downarrow{\scriptstyle f_{\bar\sigma\tau}} \\
H_{\bar\sigma}(\bar{N}) & = & H_\sigma(N) & = & H_{\bar\sigma\tau}(\bar{M})
\end{array}
$$

is commutative.

The diagram (4.7.2) shows that, if we define the image of g_M in $\mathrm{Hom}(H_{\bar\sigma}(\bar{M},R)\ ,\ H_{\bar\sigma\tau}(\bar{M}R))$ by the upper line of (4.7.2), we get elements which are functorial in $\bar{M}$ and R , and compatible with tensor products. These define elements in $Hom^{\otimes}(H_{\bar{\sigma}} \otimes R, H_{\bar{\sigma}\tau} \otimes R)$, as any object in $\underline{MM}_{\bar{k}}$ is a <u>direct factor</u> of an object $\bar{M}$ for M in $\underline{MM}_k$, see 2.20 e) .

So we have defined a map $(M\pi)^{-1}(\tau) \to \underline{Hom}^{\otimes}(H_{\bar{\sigma}}, H_{\bar{\sigma}\tau})$, which is bijective as one may see by reversing the construction, looking at (4.7.2) again.

f) Msp_1 can be defined like sp_1 in [DMOS] II 6.23 (d), but it is shorter for us here, just to define it by

$$Msp_1 = \psi \circ sp_1 .$$

g) We only have to note that $M \mapsto s.s.M$ is a tensor functor, maps $\underline{MM}_k$ to $\underline{M}_k$ by 4.5, and that for any $H \in \underline{R}_k$ there is a <u>unique</u> isomorphism

$$H \cong \bigoplus_{m \in \mathbb{Z}} Gr_m^W H$$

inducing the identity on the graded pieces, as $Hom(H,H') = 0$ for pure realizations of different weights. With respect to this isomorphism,

$$\underline{M}_k \hookrightarrow \underline{MM}_k \xrightarrow{s.s.} \underline{M}_k$$

is the identity, and s.s. commutes with H_σ . Therefore s.s. induces the homomorphism Σ and we have $\psi \circ \Sigma = id$.
As s.s. commutes with the inclusions $\underline{M}_k^o \hookrightarrow \underline{MM}_k$ and $\underline{M}_k^o \hookrightarrow \underline{M}_k$, we have $M\pi \circ \Sigma = \pi$, and the rest is clear.

<u>4.8.</u> For the description of $U(\sigma)$ we can use Saavedra's results [SR] IV §2 on filtered Tannakian categories. Namely the fibre functors H_σ on $\underline{MM}_k$ are filtered by the <u>weight filtration</u> $W_m H_\sigma = H_\sigma W_m$, and if

$$\underline{Aut}^{\otimes !}(H_\sigma)$$

is the subfunctor of $\underline{Aut}^{\otimes}(H_\sigma)$ formed by those tensor automorphisms of H_σ which induce the identity on $\mathrm{Gr}\, H_\sigma = \bigoplus_{m\in\mathbb{Z}} \mathrm{Gr}^W_m H_\sigma$, $\mathrm{Gr}^W_m H_\sigma = W_m H_\sigma / W_{m-1} H_\sigma$, then we have

<u>4.9. Proposition</u> $U(\sigma) = \underline{\mathrm{Aut}}^{\otimes !}(H_\sigma|_{\underline{MM}_k}) = \underline{\mathrm{Aut}}^{\otimes !}(H_{\bar\sigma}|_{\underline{MM}_{\bar k}})$ for $\sigma: k \hookrightarrow \mathbb{C}$, respectively $\bar\sigma: \bar k \hookrightarrow \mathbb{C}$ with $\bar\sigma|_{\bar k} = \sigma$.

<u>Proof</u> From the proof of 4.7 g) it is easy to see that $G(\sigma)$ is canonically isomorphic to the automorphism group of the fibre functor $\mathrm{Gr}\, H_\sigma = H_\sigma \cdot \mathrm{s.s.}$ on $\underline{MM}_k$. With this identification we have $U(\sigma) = \mathrm{Ker}(MG(\sigma) \to G(\sigma)) = \mathrm{Ker}(\underline{\mathrm{Aut}}^{\otimes}(H_\sigma|_{\underline{MM}_k}) \to \underline{\mathrm{Aut}}^{\otimes}(\mathrm{Gr}H_\sigma|_{\underline{MM}_k}))$ $= \mathrm{Aut}^{\otimes !}(H_\sigma|_{\underline{MM}_k})$.

The same considerations apply to $\bar k$, $MG^O(\sigma)$ and $G^O(\sigma)$ by the diagram in 4.7 c) .

<u>4.10.</u> It is often inconvenient to restrict to projective or quasi-projective varieties, and we will show that this is in fact not necessary.

Let $\underline{W}_k$ and $\overset{\circ}{\underline{W}}_k$ be the categories of smooth separated and smooth proper varieties over k , respectively . Then we can define the functors

$$H: \underline{W}_k \to \underline{R}_k$$
$$H: \overset{\circ}{\underline{W}}_k \to \underline{MR}_k$$

in exactly the same way as in section 3, because nowhere the quasi-projectivity was used. Of course, with the notations of 3, the varieties $X, Y, Y_j, Y^{(q)}$ are only smooth and proper and not necessarily projective then; this corresponds to Deligne's construction of mixed Hodge structures for smooth varieties [D4] . We now claim that we do not get new mixed motives or motives by this.

4.11. Proposition $H : \underline{W}_k \to \underline{R}_k$ factorizes through $\underline{M}_k$ and $H : \overset{\circ}{\underline{W}}_k \to \underline{MR}_k$ factorizes through $\underline{MM}_k$.

Proof Let U_o be a smooth variety over k. By Nagata [N], U_o is an open subvariety of a proper variety X_o, and by Hironaka's result on resolution of singularities we can assume that X_o is smooth. By Chow's lemma there is a projective variety X and a proper birational morphism $f: X \to X_o$, and again by Hironaka we may assume X to be smooth. Let $U = f^{-1}(U_o)$. Then $f: U \to U_o$ is proper and birational and therefore the induced map

$$f^* : H(U_o) \to H(U)$$

of mixed realizations is injective - in fact, in all three cohomology theories there is a left inverse by the transpose under Poincaré duality of the corresponding map for the cohomology with compact support. So f^* identifies $H(U_o)$ with a subobject of $H(U)$, i.e., with a mixed motive by 4.2. For U_o smooth and proper we have $U_o = X_o$ and $U = X$ and can use 4.4 d), we can also conclude by 4.4. a) as $H(X_o) \in \underline{R}_k$.

PART II

ALGEBRAIC CYCLES, K-THEORY, AND EXTENSION CLASSES

§5. The conjectures of Hodge and Tate for smooth varieties

The common object of the conjectures of Hodge and Tate is the description of the group of algebraic cycles in the cohomology of a smooth projective variety. To recall these conjectures, let k be a field with algebraic closure $\bar{k}$, $G_k = \mathrm{Gal}(\bar{k}/k)$, X a smooth projective variety over k, $\bar{X} = X \times_k \bar{k}$, and $CH^r(X)$ the Chow groups of algebraic cycles of codimension r on X modulo linear equivalence (see, e.g., [Kl]§2).

5.1. There is a canonical cycle map for $\ell \neq \mathrm{char}\, k$

$$cl_\ell^r = cl_\ell^{r,X} : CH^r(X) \longrightarrow H^{2r}_{\acute{e}t}(\bar{X}, \mathbb{Q}_\ell(r)) = H_\ell^{2r}(X)(r) \ ,$$

whose image lies in the fixed part

$$H^{2r}_{\acute{e}t}(\bar{X}, \mathbb{Q}_\ell(r))^{G_k} =: \Gamma_\ell(H_\ell^{2r}(X)(r))$$

under G_k. Tate conjectures that the image of cl_ℓ^r generates this group over $\mathbb{Q}_\ell$, if k is finitely generated as a field ([T 1]).

5.2. For $k = \mathbb{C}$ there is a cycle map

$$cl^r = cl^{r,X} : CH^r(X) \longrightarrow H^{2r}(X(\mathbb{C}), \mathbb{Q}) \ ,$$

whose image consists of (r,r)-classes, i.e., is contained in

$$H^{2r}(X(\mathbb{C}), \mathbb{Q}) \cap H^{r,r}(X, \mathbb{C}) = H^{2r}(X(\mathbb{C}), \mathbb{Q}) \cap F^r H^{2r}(X, \mathbb{C}) \ .$$

The Hodge conjecture states that the image of cl^r generates this group over $\mathbb{Q}$ (cf. [Gr]).

5.3. In our setting it is better to renormalize the last cycle map by powers of $2\pi i$ and regard it as a map

$$cl_B^r = cl_B^{r,X} : CH^r(X) \longrightarrow H^{2r}(X(\mathbb{C}), \mathbb{Q}\cdot(2\pi i)^r) = H_B^{2r}(X)(r)$$

into the r-fold Tate twist of the $\mathbb{Q}$-Hodge structure $H_B^{2r}(X) := H^{2r}(X(\mathbb{C}),\mathbb{Q})$, whose image consists of (0,0)-classes. (Note the formula $F^0(H(r))\otimes\mathbb{C} = F^rH\otimes\mathbb{C}$ for a Hodge structure H). If one works with Chern classes, this amounts to using the more natural first Chern class

$$cl_B^1 = c_1 : \mathrm{Pic}(X) = H^1(X,\mathcal{O}_X^\times) \longrightarrow H^2(X,\mathbb{Z}\cdot 2\pi i) ,$$

which is the connecting morphism for the exact sequence of analytic sheaves

$$0 \longrightarrow \mathbb{Z}\cdot 2\pi i \longrightarrow \mathcal{O} \xrightarrow{\exp} \mathcal{O}^\times \longrightarrow 0 .$$

Then no choice of $i = \sqrt{-1}$ is involved, and moreover the cycle maps cl_l^r and cl_B^r are compatible under the comparison isomorphisms between complex and étale cohomology.

Finally, for any field k of characteristic zero there is a cycle map

$$cl_{DR}^r = cl_{DR}^{r,X} : CH^r(X) \longrightarrow H_{DR}^{2r}(X/k)(r) = H_{DR}^{2r}(X)(r)$$

whose image lies in

$$F^0(H_{DR}^{2r}(X)(r)) = (F^rH_{DR}^{2r}(X))(r) ,$$

and which for any $\sigma : k \hookrightarrow \mathbb{C}$ is compatible with the map

$$cl_\sigma^r = cl_\sigma^{r,X} : CH^r(X) \xrightarrow{\sigma^*} CH^r(\sigma X) \xrightarrow{cl_B^{r,\sigma X}} H^{2r}(\sigma X(\mathbb{C}),\mathbb{Q})(r) = H_\sigma^{2r}(X)(r)$$

under the comparison isomorphism $I_{\infty,\sigma}^{2r}(X)(r)$.

5.4. Thus for $\operatorname{char} k = 0$ we obtain a cycle map

$$cl_{AH}^r = cl_{AH}^{r,X} : CH^r(X) \longrightarrow \Gamma_{AH}(H^{2r}(X)(r))$$

into the group of absolute Hodge cycles in $H^{2r}(X)(r)$ (denoted $\Gamma(H^{2r}(X)(r))$ in the first part). As a combination and weaker form of 5.1 and 5.2 one may conjecture that the image of cl^r generates the $\mathbb{Q}$-vector space $\Gamma_{AH}(H^{2r}(X)(r))$. In fact, by the inclusions

$$\begin{array}{ccc} H_\sigma^{r,r}(X) \cap H_\sigma(X)(r) & & H_\ell^{2r}(X)(r)^{G_k} \\ \cup\!\!| & & \subseteq \\ & \Gamma_{AH}(H^{2r}(X)(r)) & \\ & \cup\!\!| & \\ & \Gamma_{alg}(H^{2r}(X)(r)) := \operatorname{Im} cl^{r,X} \otimes \mathbb{Q} & \end{array} \tag{5.4.1}$$

we see, that this is implied by either the Hodge or the Tate conjecture. More precisely, we have

5.5. Lemma a) Let $k_0 \subset k$ be a finitely generated field such that X is defined over k_0 . If Tate's conjecture is true for X and every finite extension of k_0 , then conjecture 5.4 is true for X and k .

b) If the Hodge conjecture is true for σX for some embedding $\sigma : K \hookrightarrow \mathbb{C}$, then conjecture 5.4 is true for X/k .

Proof a) It is clear that every absolute Hodge cycle over $\bar{k}$ is defined over some finitely generated extension of $\bar{k}_0$, since this is true for every element in $H_{DR}(\bar{X}/\bar{k}) = H_{DR}(\bar{X}_0/\bar{k}_0) \otimes_{\bar{k}_0} \bar{k}$. By 2.19 it is therefore defined over some finitely generated extension of k_0 , i.e., it suffices to consider the case that k is finitely generated over k_0 . It is then proved in [DMOS]I 2.9 that the absolute Hodge cycles over $\bar{k}_0$ and $\bar{k}$ are the same. By assumption and 5.4.1, all absolute Hodge cycles over $\bar{k}_0$ are algebraic, hence the same statement for $\bar{k}$, and for k by taking fixed modules under G_k , see 2.19.

b) It suffices to consider k algebraically closed. By assumption and 5.4.1, every absolute Hodge cycle is algebraic over some field k' which is finitely generated over k , and we get an algebraic cycle over k by specialization, see [DV] exp. 0 .

5.6. We note that the above conjectures would imply some other weaker ones:

a) $\Gamma_{AH}(H^{2r}(X)(r)) \hookrightarrow H_\sigma^{r,r}(X) \cap H_\sigma^{2r}(X)(r)$ should be surjective for $k = \bar{k}$ and $\sigma : k \hookrightarrow \mathbb{C}$; this is Deligne's "espoire" that every Hodge cycle is absolute Hodge see [D6].

b) $\Gamma_{AH}(H^{2r}(X)(r)) \otimes \mathbb{Q}_\ell \hookrightarrow H_\ell^{2r}(X)(r)^{G_k}$ should be surjective for k finitely generated over the prime field.

c) $I^{2r}_{\ell,\bar{\sigma}}(r) : H_\sigma^{2r}(X) \otimes \mathbb{Q}_\ell \xrightarrow{\sim} H_\ell^{2r}(X)(r)$ for $\bar{\sigma} : \bar{k} \hookrightarrow \mathbb{C}$ with $\bar{\sigma}|_k = \sigma$ should induce an isomorphism

$$[H_\sigma^{r,r}(X) \cap H_\sigma^{2r}(X)(r)] \otimes \mathbb{Q}_\ell \simeq H_\ell^{2r}(X)(r)^{G_k}$$

for k finitely generated and sufficiently big. In general, no "inclusion" is known, but a) would imply " $\subseteq$ " and b) would imply " $\supseteq$ " . In the first case we could conclude Tate $\Rightarrow$ Hodge, in the second case we had Hodge $\Rightarrow$ Tate.

d) $\dim_{\mathbb{Q}}(H_{\sigma}^{r,r}(X) \cap H_{\sigma}^{2r}(X)(r)$ should be independent of σ.

e) $\dim_{\mathbb{Q}_\ell}[H_{\ell}^{2r}(X)(r)]^{G_k}$ should be independent of ℓ.

f) These dimensions should be equal for k finitely generated and sufficiently big.

5.7. Remark Deligne has proved a) for abelian varieties X, see [DMOS]I 2.11; and a)-c) hold for abelian varieties with complex multiplication by the work of Shimura-Taniyama and Serre, even in the stronger form stated in [Se 2] § 3, compare also Pohlmann [P]. The results can be extended to the category of motives generated by abelian varieties, containing for example K3-surfaces and Fermat hypersurfaces, compare [DMOS] 6.26 and 6.27.

In fact, conjectures a)-c) have convenient interpretations in the setting of the associated Tannakian categories. For example, let $MT(H_{\sigma}^{2r}(X))$ be the Mumford-Tate group of the Hodge structure $H_{\sigma}^{2r}(X)$. It is the subgroup of $GL_{\mathbb{Q}}(H_{\sigma}^{2r}(X))$ fixing all Hodge cycles in all products $H_{\sigma}^{2r}(X)^{\otimes s} \otimes (H_{\sigma}^{2r}(X)^{\vee})^{\otimes t} \otimes \mathbb{Q}(1)^{\otimes u}$ for all $s,t \in \mathbb{N}_0$, $u \in \mathbb{Z}$, and thus the "Galois group" of the Tannakian category generated by the Hodge structures $H_{\sigma}^{2r}(X)$ and $\mathbb{Q}(1)$, see [DMOS] I § 3.

On the other hand, let $G(H^{2r}(X),\sigma)$ be the "Galois group" of the Tannakian subcategory of $\underline{M}_k$ generated by $H^{2r}(X)$ and $1(1)$ (with fibre functor H_{σ}), i.e., the image of $G(\sigma) \longrightarrow GL_{\mathbb{Q}}(H_{\sigma}^{2r}(X))$. Then by (5.4.1) for all tensors we have

$$MT(H_{\sigma}^{2r}(X)) \leq G(H^{2r}(X),\sigma) \ ;$$

and a) for all tensors would imply equality.

b) and c) have similar interpretations. However, one does not know in general, whether G_k acts semisimply on $H_{\ell}^{2r}(X)$ as conjectured by Grothendieck and Serre, so one also has to consider subquotients of the tensor products for $H^{2r}_{\ell}(X)$ ([DMOS] I. 3.2). In any case, c) is related to the conjecture that $\mathrm{Im}(G_k \longrightarrow GL_{\mathbb{Q}_\ell}(H^i_\ell(X)))$ and $MT(H^i_\sigma(X))(\mathbb{Q}_\ell)$ are commensurable, see [Se 2]§ 3.

5.8. Similar conjectures for non-proper varieties seem to give nothing new, at least for k of characteristic zero. In fact, let U be a smooth quasi-projective variety over k and let X be a smooth projective compactification. Then there are cycle maps

$$cl^r_{AH} = cl^{r,U}_{AH} : CH^r(U) \longrightarrow \Gamma_{AH}(H^{2r}(U)(r))$$

as before, having components cl^r_{DR}, cl^r_ℓ and cl^r_σ with images in

$$\Gamma_{DR}(H^{2r}_{DR}(U)(r)) = W_0(H^{2r}_{DR}(U)(r)) \cap F^0(H^{2r}_{DR}(U)(r) = W_{2r}H^{2r}_{DR}(U) \cap F^r H^{2r}_{DR}(U) ,$$

$$\Gamma_\ell(H^{2r}_\ell(U)(r)) = H^{2r}_\ell(U)(r)^{G_k} \cap W_0(H^{2r}_\ell(U)(r)) ,$$

$$\Gamma_H(H^{2r}_\sigma(U)(r)) = W_0(H^{2r}_\sigma(U)(r)) \cap F^0(H^{2r}_\sigma(U)(r) \otimes \mathbb{C})$$

$$(5.8.1) \qquad = (2\pi i)^r W_{2r}H^{2r}(\sigma U,\mathbb{Q}) \cap F^r H^{2r}(\sigma U,\mathbb{C}) ,$$

respectively (for example by using fundamental classes in the relative cohomology $H^{2r}_Z(U)$ for a prime cycle Z of codimension r). But then $cl^{r,U}_?$ factorizes through

$$\Gamma_?(W_0H^{2r}_?(U)(r)) = \Gamma_?(\mathrm{Im}\ H^{2r}_?(X)(r) \longrightarrow H^{2r}_?(U)(r))$$

(see 3.22, or [D4] 3.2.17, for the last equality), and by lemma 1.1, $W_0H^{2r}(U)(r)$ is a direct factor of $H^{2r}(X)(r)$, so that the maps

$$\Gamma_?\ H^{2r}_?(X)(r) \longrightarrow \Gamma_?\ H^{2r}_?(U)(r)$$

are surjective. On the other hand, the restriction $CH^r(X) \longrightarrow CH^r(U)$

is also surjective, and the diagram

$$\begin{array}{ccc} CH^r(X) & \xrightarrow{cl_?^{r,X}} & \Gamma_? \, H_?^{2r}(X)(r) \\ \downarrow & & \downarrow \\ CH^r(U) & \xrightarrow{cl_?^{r,U}} & \Gamma_? \, H_?^{2r}(U)(r) \end{array}$$

commutes. This shows the following.

The only possible formulation of the conjectures of Hodge and of Tate for U is that for $k = \mathbb{C}$ $\Gamma_H(H_{id}^{2r}(U)(r)) = F^rH^{2r}(U,\mathbb{C}) \cap W_{2r}H^{2r}(U,\mathbb{Q})$ and for finitely generated k $\Gamma_\ell(H_\ell^{2r}(U)(r)) = H_{et}^{2r}(U \times_k \bar{k},\mathbb{Q}_\ell(r))^{G_k}$ should be generated by algebraic cycles. At the same time, these conjectures are immediately implied by those for X.

The same holds for conjecture 5.4 involving $\Gamma_{AH}(H^{2r}(U)(r))$, and also for the Tate conjecture in characteristic $p > 0$, if one has resolution of singularities and semi-simple action of G_k on $H_\ell^{2r}(X)$ (this will be discussed more generally in § 7). So again, at least morally, we obtain nothing new.

5.9. However, for smooth non-proper varieties U/k it makes sense to study the space

$$\Gamma_{AH}(H^i(U)(j))$$

and those defined in 5.8.1 for arbitrary $i,j \in \mathbb{Z}$. For X smooth and proper over k, $\Gamma(H^i(X)(j))$ is zero unless $i = 2j$, because $H^i(X)(j)$ is pure of weight $i-2j$ and in general

$$\Gamma(H) = \Gamma(W_0H) \hookrightarrow \Gamma(Gr_0^W H) \tag{5.9.1}$$

for any mixed realization H. But otherwise the space above can be non-zero for $i \leq 2j$, and is indeed connected with interesting questions.

For example, if X is a smooth projective curve over k and $x \neq y$ are two k-rational points, we get an exact sequence for $U = X \setminus \{x,y\}$

$$0 \longrightarrow H^1(X) \longrightarrow H^1(U) \longrightarrow H^0(\{x,y\})(-1) \xrightarrow{\delta} H^2(X) \longrightarrow 0 ,$$

in which $\delta(1)$ factorizes through the cycle map $cl^{1,X}$. Therefore we get an exact sequence

$$(5.9.2) \qquad \begin{array}{ccccccccc} 0 & \longrightarrow & H^1(X) & \longrightarrow & H^1(U) & \longrightarrow & 1(-1) & \longrightarrow & 0 , \\ & & \text{weight } 1 & & & & \text{weight } 2 & & \end{array}$$

where $1(-1)$ has a "basis" $1_x - 1_y$. It is a non-trivial question whether this sequence splits or not: think of the extension of Galois representations in the l-adic realization and of the mixed Hodge structure of $H^1(U)$ in the Hodge realization. Since kernel and cokernel have different weights, a section of 5.9.2 is given by a non-trivial element in

$$\mathrm{Hom}(1(-1),H^1(U)) = \mathrm{Hom}(1,H^1(U)(1)) = \Gamma(H^1(U)(1)) ,$$

and we shall show in § 9 that there is such a section if and only if $(x)-(y)$ is a torsion point in the Jacobian of X.

This suggests to look for "algebraic elements" in $\Gamma(H^i(U)(j))$, and there are indeed some, given by higher algebraic K-theory. First recall, that the rational cycle maps cl^r can also be described by Chern characters

$$ch_j : K_0(U) \longrightarrow \Gamma_?(H^{2j}(U)(j))$$

on the Grothendieck group of locally free $\mathcal{O}_U$-modules, via the isomorphism

$$K_0(U) \otimes \mathbb{Q} \cong \bigoplus_{i\geq 0} Gr^i_\gamma K_0(U) \otimes \mathbb{Q} \cong \bigoplus_{i\geq 0} CH^i(U) \otimes \mathbb{Q} ,$$

where $\gamma^i K_0(U)$ is the γ-filtration [SGA 6] exp. 0. Now by results of Schechtman [Sche] and Gillet [Gi] generalizing earlier work of Soulé [Sou 1] there are higher Chern characters

$$(5.9.3) \qquad ch_{i,j} : K_{2j-1}(U) \longrightarrow \Gamma_?(H^i(U)(j))$$

on Quillen's higher K-groups such that $ch_{2j,j}$ coincides with ch_j above. In the cited references the $ch_{i,j}$ are defined for the singular, the étale and the de Rham cohomology; to get a morphism into the group of absolute Hodge cycles one has to check that these are compatible under the comparison isomorphisms. But the Chern characters are defined by means of Chern classes

$$c_{i,j} : K_{2j-1}(U) \longrightarrow \Gamma_?(H^i(U)(j)) ,$$

so by reducing to universal Chern classes and using the splitting principle one only has to show the compatibility for the first Chern class $ch_{2,1} = ch_1 = cl^1 : Pic(U) \longrightarrow \Gamma_?(H^2(U)(1))$, which we already used in 5.8.

For a generalization of the Hodge and the Tate conjecture we propose to study the image of the maps 5.9.3 for general i and j. First we investigate for which i and j the target groups can be non-zero, by studying the weights of the realizations. Namely, for each realization H - in the sense of § 3, or an ℓ-adic one, or a Hodge structure, or a de Rham realization - we have a weight filtration, and say that the weight $w \in \mathbb{Z}$ occurs in H, if $Gr_w^W H \neq 0$.

<u>5.10. Lemma</u> Let U be a smooth variety of dimension d over k. Then for the weights w occuring in $H^i(U)(j)$ we have

$$i-2j \leq w \leq 2i-2j \text{ , if } 0 \leq i \leq d$$

$$i-2j \leq w \leq 2d-2j \text{ , if } d \leq i \leq 2d \text{ .}$$

<u>Proof</u> See [D4] 3.2.15 b) and [D9] 3.3.8.

<u>5.11. Corollary</u> One has $\Gamma(H^i(U)(j)) \neq 0$ at most for

$$0 \leq j \leq d \text{ and } j \leq i \leq 2j \text{ .}$$

<u>Proof</u> In view of 5.9.1 we must have

$$i-2j \leq 0 \leq 2i-2j \text{ and } 0 \leq i \leq d$$

or $\quad i-2j \leq 0 \leq 2d-2j$ and $d \leq i \leq 2d$.

(5.11.1)

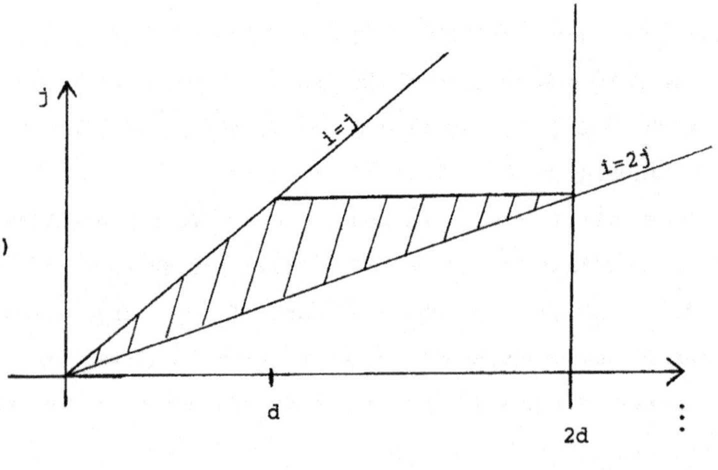

5.12. For the study of the maps 5.9.3 it is convenient to consider the action of the Adams operators ψ^k, $k \geq 1$, on the K-groups $K_m(U)$ (see [Sou 3] for this and the following). If we set

$$K_m(U)^{(j)} = \{x \in K_m(U) \otimes_{\mathbb{Z}} \mathbb{Q} \mid \psi^k(x) = k^j \cdot x \quad \text{for all} \quad k \in \mathbb{N}\} ,$$

then $K_m(U) \otimes \mathbb{Q} = \bigoplus_{j \geq 0} K_m(U)^{(j)}$ and $K_m(U)^{(j)}$ is canonically isomorphic to the graded term $Gr^i_\gamma K_m(U) \otimes \mathbb{Q}$ for the γ-filtration. Since $ch_{i,j}(\psi^k(x)) = k^j ch_{i,j}(x)$, the map $ch_{i,j}$ vanishes on $K_{2j-i}(U)^{(\nu)}$ for $\nu \neq j$ and it suffices to consider the restriction

$$ch_{i,j} : K_{2j-1}(U)^{(j)} \longrightarrow \Gamma_?(H^i(U)(j)) .$$

Following Beilinson [Bei 2] we define the "motivic cohomology" of U by

$$H^i_{\mathcal{M}}(U,\mathbb{Q}(j)) := K_{2j-i}(U)^{(j)}$$

(denoted "absolute cohomology" $H^i_A(U,\mathbb{Q}(j))$ in [Bei 1]), so that we study the morphisms

$$ch_{i,j} : H^i_{\mathcal{M}}(U,\mathbb{Q}(j)) \longrightarrow \Gamma_?(H^i(U)(j))$$

from the motivic cohomology to the various other cohomology theories.

We can describe their image in the following case.

5.13. Theorem Let U be a smooth connected variety over $\mathbb{C}$, then the Chern class induces an isomorphism

$$c_{1,1} : \mathcal{O}(U)^\times/\mathbb{C}^\times \xrightarrow{\sim} \Gamma_{\mathcal{H}}(H^1_B(U,\mathbb{Z})(1)) = 2\pi i \cdot W_2H^1(U,\mathbb{Z}) \cap F^1H^1(U,\mathbb{C}) ,$$

in particular,

$$ch_{1,1} \; : \; K_1(U)^{(1)} \longrightarrow \Gamma_H(H^1_B(U)(1))$$

is surjective.

<u>Proof</u> We shall get two proofs of this fact. The one given here uses the Beilinson-Deligne cohomology $H^i_{\mathcal{D}}(U,\mathbb{Z}(j))$ of U (see [Bei 1], [EV]); the second one will be given in 9.11, is based on the theorem of Abel-Jacobi, and shows the relation with (5.9.2).

It is shown in [EV] that there is an isomorphism

$$\mathcal{O}(U)^\times \underset{\sim}{\xrightarrow{\alpha}} H^1_{\mathcal{D}}(U,\mathbb{Z}(1)) \;,$$

so the claim follows from the commutative diagrams

$$\begin{array}{ccccccccc} 0 & \longrightarrow & \mathbb{C}/\mathbb{Z}(1) & \longrightarrow & H^1_{\mathcal{D}}(U,\mathbb{Z}(1)) & \longrightarrow & \Gamma_H(H^1_B(U,\mathbb{Z})(1)) & \longrightarrow & 0 \\ & & \uparrow & & \uparrow & & \uparrow & & \\ 0 & \longrightarrow & \mathbb{C}^\times & \longrightarrow & \mathcal{O}(U)^\times & \longrightarrow & \mathcal{O}(U)^\times/\mathbb{C}^\times & \longrightarrow & 0 \end{array}$$

and

$$\begin{array}{ccc} & & \mathcal{O}(U)^\times \\ & \overset{\det}{\nearrow} & \wr\downarrow\alpha \\ K_1(U) & \xrightarrow{\quad c_{1,1}\quad} & H^1_{\mathcal{D}}(U,\mathbb{Z}(1)) \\ & \underset{c_{1,1}}{\searrow} & \downarrow \\ & & \Gamma_H(H^1_B(U,\mathbb{Z}(1))) \end{array}$$

together with the fact that det induces an isomorphism

$$K_1(U)^{(1)} \underset{\sim}{\longrightarrow} \mathcal{O}(U)^\times \otimes \mathbb{Q} \;,$$

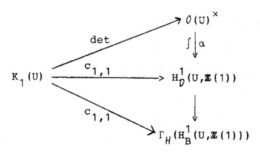

compare [Sou 3].

5.14. Remark The proof above is very similar to the proof of the Hodge conjecture for divisors by using the exponential sequence, which can be reinterpreted as a quasi-isomorphism $\mathbb{Z}(1)_{\mathcal{D}} \xrightarrow{\sim} \mathbb{G}_m[-1]$ for a smooth proper variety. However, for non-proper U as above this quasi-isomorphism holds true no longer, so we really have to use the Beilinson-Deligne cohomology instead of the exponential sequence.

The following generalizes a result of Friedlander [Fr] Prop. 3.6.

5.15. Theorem Let U be a smooth, geometrically connected variety over a finitely generated field k . and let ℓ be a prime, $l \neq \text{char}\, k$. Then the connecting morphism for the Kummer sequences

$$0 \longrightarrow \mu_{\ell^n} \longrightarrow \mathbb{G}_m \xrightarrow{\ell^n} \mathbb{G}_m \longrightarrow 0$$

induce isomorphisms

$$\text{a)}\quad (\mathcal{O}(U)^\times)^\wedge = \varprojlim_n \mathcal{O}(U)^\times/\ell^n \underset{\sim}{\xrightarrow{\delta}} \varprojlim_n H^1_{et}(U,\mu_{\ell^n}) = H^1(U,\mathbb{Z}_\ell(1))$$

and

$$\text{b)}\quad (\mathcal{O}(U)^\times/k^\times)^\wedge \cong (\mathcal{O}(U)^\times/k^\times) \otimes \mathbb{Z}_\ell \xrightarrow{\sim} H^1(\bar{U},\mathbb{Z}_\ell(1))^{G_k},$$

in particular, the first Chern class

$$c_{1,1} : K_1(U) \otimes \mathbb{Z}_\ell \longrightarrow H^1(\bar{U},\mathbb{Z}_\ell(1))^{G_k}$$

is surjective.

Proof a) follows by passing to the inverse limit over the exact sequences

$$0 \longrightarrow \mathcal{O}(U)^{\times}/\ell^n \longrightarrow H^1_{\text{ét}}(U,\mu_{\ell^n}) \longrightarrow {}_{\ell^n}\mathrm{Pic}(U) \longrightarrow 0\ ,$$

since Pic(U) is finitely generated by the generalized Mordell-Weil theorem for finitely generated k, cf. [La] II 7.6. For b) we use continuous cohomology and the five term exact sequence

$$0 \to H^1_{\text{cont}}(G_k,\mathbb{Z}_\ell(1)) \to H^1_{\text{cont}}(U,\mathbb{Z}_\ell(1)) \xrightarrow{\text{res}} H^1(\bar{U},\mathbb{Z}_\ell(1))^{G_k} \to H^2_{\text{cont}}(G_k,\mathbb{Z}_\ell(1)) \xrightarrow{\pi^*} H^2_{\text{cont}}(U,\mathbb{Z}_\ell(1))$$

induced by the Hochschild-Serre spectral sequence [J1] 3.5. Here π is induced by the morphism $\pi : U \to \mathrm{Spec}\ k$. If U has a k-rational point, π has a section and hence π^* is injective. In general, let K/k be a Galois extension with Galois group G . Then we have a commutative exact diagram

$$\begin{array}{ccccc}
H^2(G,\mathbb{Z}_\ell(1)^{G_K}) = 0 & & & & \\
\uparrow & & & & \\
H^1_{\text{cont}}(U\times K,\mathbb{Z}_\ell(1))^G & \xrightarrow{\text{res}} & (H^1(\bar{U},\mathbb{Z}_\ell(1))^{G_K})^G & \longrightarrow & H^1(G,H^1_{\text{cont}}(G_K,\mathbb{Z}_\ell(1))) \\
\uparrow & & \| & & \\
H^1_{\text{cont}}(U,\mathbb{Z}_\ell(1)) & \xrightarrow{\text{res}} & H^1(\bar{U},\mathbb{Z}_\ell(1))^{G_k} & & \\
\uparrow & & & & \\
H^1(G,\mathbb{Z}_\ell(1)^{G_K}) = 0\ . & & & &
\end{array}$$

Since $H^1_{\text{cont}}(G_K,\mathbb{Z}_\ell(1)) = \varprojlim_n H^1(G_K,\mu_{\ell^n}) \cong \varprojlim_n K^{\times}/(K^{\times})^{\ell^n} =: \hat{K}^{\times}$ by Kummer theory, and $\hat{K}^{\times}/(K^{\times}\otimes\mathbb{Z}_\ell)$ is uniquely divisible,

$H^1(G, H^1_{cont}(G_K, \mathbb{Z}_\ell(1))) \cong H^1(G, K^\times) \otimes \mathbb{Z}_\ell = 0$, so both restrictions are surjective.

We get a commutative exact diagram

$$\begin{array}{ccccccccc} 0 & \longrightarrow & H^1_{cont}(G_k, \mathbb{Z}_\ell(1)) & \longrightarrow & H^1_{cont}(U, \mathbb{Z}_\ell(1)) & \xrightarrow{res} & H^1(\bar{U}, \mathbb{Z}_\ell(1))^{G_k} & \longrightarrow & 0 \\ (5.15.1) & & \uparrow \delta & & \uparrow \delta & & \uparrow & & \\ 0 & \longrightarrow & k^\times & \longrightarrow & \mathcal{O}(U)^\times & \longrightarrow & A & \longrightarrow & 0 \; , \end{array}$$

in which $A = \mathcal{O}(U)^\times / k^\times$ is finitely generated and torsion free. This can be seen from the diagram

$$\begin{array}{ccccccc} 0 & \longrightarrow & k^\times & \longrightarrow & \mathcal{O}(U)^\times & \longrightarrow & \bigoplus_{x \in X^{(1)} \setminus U^{(1)}} \mathbb{Z} \\ & & \| & & \cap\downarrow & & \cap\downarrow \\ 0 & \longrightarrow & k^\times & \longrightarrow & k(U)^\times & \longrightarrow & \bigoplus_{x \in X^{(1)}} \mathbb{Z} \\ & & & & d \downarrow & & \downarrow pr \\ & & & & \bigoplus_{x \in U^{(1)}} \mathbb{Z} & = & \bigoplus_{x \in U^{(1)}} \mathbb{Z} \end{array}$$

for a normal compactification X of U. Here $k(U)$ is the function field of U (and of X), $U^{(1)}$ and $X^{(1)}$ are the sets of points of codimension 1 in U and X, respectively, and d is the differential of the Quillen spectral sequence ([Q1] 5.4), i.e., $d(f) = \sum v_x(f)$ where v_x is the valuation at $x \in X^{(1)}$ or $U^{(1)}$.

Hence $\hat{A} = A \otimes \mathbb{Z}_\ell$, and we get b) by passing to the ℓ-completion in 5.14.1. The rest follows from the commutative diagram

$$\begin{array}{ccc} & \overset{det}{\nearrow\!\!\!\!\!\nearrow} & \mathcal{O}(U)^\times \\ K_1(U) & & \downarrow res \circ \delta \\ & \underset{c_{1,1}}{\searrow} & H^1(\bar{U}, \mathbb{Z}_\ell(1))^{G_k} \end{array} ,$$

note however, that $\mathcal{O}(U)^{\times} \otimes \mathbb{Z}_\ell \longrightarrow (\mathcal{O}(U)^{\times})^\wedge$ will not be an isomorphism unless k is a finite field.

5.16. Remark a) Like for 5.13 we used a suitable "absolute" cohomology theory for the proof above and shall get another proof in § 9, related to extension classes.

b) For smooth U, not necessarily geometrically connected, 5.15 a) remains true without change, and instead of b) we have

$$(\mathcal{O}(U)^{\times}/\bigoplus_{x\in U^{(0)}} k_x^{\times})^\wedge \cong (\mathcal{O}(U)^{\times}/\bigoplus_{x\in U^{(0)}} k_x^{\times}) \otimes \mathbb{Z}_\ell \xrightarrow{\sim} H^1(\bar{U},\mathbb{Z}_\ell(1))^{G_k},$$

where k_x is the separabel closure of k in $\kappa(x)$. For this we may assume that U is irreducible; let $\tilde{k}$ be the separabel closure of k in the function field of U. Since $H^q(U \times_k \bar{k},\mathbb{Z}_\ell(j)) \cong \mathrm{Ind}_{G_{\tilde{k}}}^{G_k} H^q(U \times_{\tilde{k}}\bar{k},\mathbb{Z}_\ell(j))$, we have $H^p_{cont}(G_k,H^q(U \times_k\bar{k},\mathbb{Z}_\ell(j))) \cong H^p(G_{\tilde{k}},H^q(U \times_{\tilde{k}}\bar{k},\mathbb{Z}_\ell(j)))$ for all $p,q \geq 0$, so we may replace k by $\tilde{k}$ in the above considerations.

5.17. Corollary Let U be a smooth variety over a field k which is embeddable in $\mathbb{C}$. Then the map

$$ch_{1,1} : K_1(U)^{(1)} \longrightarrow \Gamma_{AH}(H^1(U)(1))$$

is surjective.

Proof First assume that k is algebraically closed. By 5.13, the map $ch_{1,1} : K_1(U \times_k \mathbb{C})^{(1)} \longrightarrow \Gamma_{AH}(H^1(U \times_k \mathbb{C})(1))$ is surjective for a fixed embedding $k \hookrightarrow \mathbb{C}$. On the other hand, every element $x \in K_1(U \times_k \mathbb{C})$ lies in the image of the restriction $K_1(U \times_k R) \longrightarrow K_1(U \times_k \mathbb{C})$ for some finitely generated k-algebra

R, $k \subseteq R \subseteq \mathbb{C}$, see [Q1] § 7, 2.2.

Choosing a closed point $\alpha_2 : R \twoheadrightarrow k \hookrightarrow \mathbb{C}$ in the same connected component as the "generic point" $\alpha_1 : R \subseteq \mathbb{C}$ we see that image of x in $\Gamma_{AH}(H^1(U \times_k \mathbb{C})(1))$ lies in the image of

$$K_1(U) \longrightarrow \Gamma_{AH}(H^1(U)(1)) \longrightarrow \Gamma_{AH}(H^1(U \times_k \mathbb{C})(1)) ,$$

since we have

$$\alpha_1^* = \alpha_2^* : \Gamma_{AH}(H^1(U \times_k R)(1)) \longrightarrow \Gamma_{AH}(H^1(U \times_k \mathbb{C})(1)) ,$$

as can be checked, for example, in the ℓ-adic realization via the Künneth formula.
If k is not algebraically closed, we may apply the trace with respect to some finite extension K/k .

I want to state and discuss the following

<u>5.18. Conjecture</u> If U is a smooth variety over a number field k , then for every $i,j \geq 0$

$$ch_{i,j} : K_{2j-i}(U) \otimes \mathbb{Q} \longrightarrow \Gamma_{AH}(H^i(U)(j))$$

is surjective.

I also think that the following "Tate version" of it should be true, replacing Γ_{AH} by Γ_ℓ .

<u>5.19. Conjecture</u> If k is a finite field or a global field and U is a smooth variety over k , then for every $\ell \neq char(k)$ and $i,j \geq 0$ the map

$$ch_{i,j} : K_{2j-i}(U) \otimes \mathbb{Q}_\ell \longrightarrow H^i_{et}(U \times_k \bar{k}, \mathbb{Q}_\ell(j))^{G_k}$$

is surjective.

In view of 5.15 and the discussion in 5.8 it is very tempting to state conjecture 5.19 (like the Tate conjecture) more generally for a finitely generated field k, but we shall show in § 9 that it becomes false in general, if k contains too many parameters. The same can be said for 5.18.

The obvious " Hodge version" of 5.18 - replacing Γ_{AH} by Γ_H for $k = \mathbb{C}$ - is contained in a conjecture stated by Beilinson in [Bei 2], but we shall see that this is false in general by the same arguments as above. I think that the following special case should be true.

<u>5.20. Conjecture</u> Let U be a smooth variety over $\mathbb{C}$ that can be defined over a number field k. Then for all $i,j \in \mathbb{Z}$ the Chern character

$$ch_{i,j} : K_{2j-i}(U) \otimes \mathbb{Q} \rightarrow (2\pi i)^j W_{2i} H^i(U,\mathbb{Q}) \cap F^j H^i(U,\mathbb{C}) = \Gamma_H(H^i_B(U)(j))$$

is surjective.

The next statement shows that we may restrict our attention to the cases $k = \mathbb{Q}$ or $k = \mathbb{F}_p(t)$ or $k = \mathbb{F}_p$ for a prime p.

<u>5.21. Lemma</u> Let K/k be a finite separabel extension. Then conjecture 5.18 (resp. 5.19, resp. 5.20) is true for k if and only if it is true for K.

<u>Proof</u> By applying this to N/K and N/k where N is the normal closure of K/k, we may consider the case that K/k is Galois with Galois group G. For a variety U over k let $U \times_k K$ be the base

extension, and for a variety V over K let $R_{K/k}$ be the Grothendieck restriction $V \longrightarrow \operatorname{Spec} K \longrightarrow \operatorname{Spec} k$. Then the claim for 5.18 follows from the commutative diagrams

$$\begin{array}{ccc} K_{2j-i}(V) & \longrightarrow & \Gamma_{AH}(H^i(V)(j)) \\ \| & & \| \\ K_{2j-i}(R_{K/k}V) & \longrightarrow & \Gamma_{AH}(H^i(R_{K/k}V)(j)) \end{array} \qquad \begin{array}{ccc} K_{2j-i}(U \times_k K)^G_{\mathbb{Q}} & \longrightarrow & \Gamma_{AH}(H^i(U \times_k K)(j))^G \\ \uparrow\wr & & \uparrow\wr \\ K_{2j-i}(U)_{\mathbb{Q}} & \longrightarrow & \Gamma_{AH}(H^i(U)(j)) \end{array},$$

following from 2.19, 2.20 and the relations $H^i(R_{K/k}V)(j) = R_{K/k}H^i(V)(j)$ and $H^i(U \times_k K)(j) = H^i(U)(j) \times_k K$. For 5.19 one uses the corresponding diagrams with Γ_{AH} replaced by Γ_ℓ, since $H^i_{et}(V \times_K \bar{k}, \mathbb{Q}_\ell(j))^{G_K} = H^i_{et}(R_{K/k}V \times_k \bar{k}, \mathbb{Q}_\ell(j))^{G_k}$ and $H^i(U \times_k \bar{k}, \mathbb{Q}_\ell(j))^{G_k} = (H^i(U \times_k K \times_K \bar{k}, \mathbb{Q}_\ell(j))^{G_K})^G$.

For 5.20 let V be a variety over $\mathbb{C}$, let V_0 be a variety over the number field K such that $V \cong V_0 \times_{K,\delta_0} \mathbb{C}$ for some embedding $\delta_0 : K \hookrightarrow \mathbb{C}$, and let $U_0 = R_{K/k}V_0$. Then the canonical $\mathbb{C}$-morphism $\psi : V \longrightarrow V = U_0 \times_{k,\delta_0} \mathbb{C} = \coprod_{\delta:K\hookrightarrow\mathbb{C}} V_0 \times_{K,\delta} \mathbb{C}$, which is the inclusion of the component $V_0 \times_{K,\delta_0} \mathbb{C}$, induces a commutative diagram

$$\begin{array}{ccc} K_{2j-1}(U) & \longrightarrow & \Gamma_H(H^i_B(U)(j)) \\ \psi_*\uparrow\ \downarrow\psi^* & & \psi_*\uparrow\ \downarrow\psi^* \\ K_{2j-i}(V) & \longrightarrow & \Gamma_H(H^i_B(V)(j)) \end{array}$$

with $\psi^*\psi_* = \mathrm{id}$. Hence, if 5.20 is true for U, it is true for V. The conclusion from k to K is trivial.

The conjectures above have some remarkable consequences, in that properties of the K-theory would imply similar ones for the realizations. The following argument is copied from Beilinson [Bei 1]; it is based on a fundamental result of Suslin:

5.22. Theorem ([Su 2][Sou 3]3) If F is a field, then

a) $H^i_{\mathcal{M}}(F,\mathbb{Q}(j)) = 0$ for $i > j$,

b) $H^i_{\mathcal{M}}(F,\mathbb{Q}(i)) \cong K_i^{\mathrm{Milnor}}(F) \otimes \mathbb{Q}$ (Milnor K-theory).

Recall that for any presheaf G for the Zariski topology on a scheme X the filtration by coniveau is defined by

$$N^iG(X) = \bigcup_{\substack{U \subseteq X \text{ open} \\ \operatorname{codim}_X(X \setminus U) \geq i}} \operatorname{Ker}(G(X) \longrightarrow G(U)) .$$

In these terms, Suslin's theorem implies:

5.23. Corollary Let U be a smooth variety over a field k, then $H^i_{\mathcal{M}}(U,\mathbb{Q}(j))$ has support in codimension $i-j$, i.e., $N^{i-j}H^i_{\mathcal{M}}(U,\mathbb{Q}(j)) = H^i_{\mathcal{M}}(U,\mathbb{Q}(j))$.

Proof By a result of Soulé ([Sou 3] théorème 4) the Quillen spectral sequence in K-theory ([Q1] 5.4) induces a spectral sequence

$$(5.23.1) \quad E_1^{p,q}(U)(j) = \bigoplus_{x \in U^{(p)}} K_{-p-q}(\kappa(x))^{(j-p)} \Rightarrow K_{-p-q}(U)^{(j)} ,$$

where $U^{(p)}$ is the set of points of codimension p of U and $\kappa(x)$ is the residue field of x. By 5.22 a) we have $E_1^{p,q}(U)(j) = 0$ for $j-p > -p-q$, For $-p-q = 2j-i$ we see that $E_1^{p,q} = 0$ for $p < i-j$, i.e., the part of 5.23.1 contributing to $H^i_{\mathcal{M}}(U,\mathbb{Q}(j))$ lives in codimension $\geq i-j$.

5.24. Conjectures 5.18 to 5.20 predict the same behaviour for

$\Gamma_{AH}(H^i(U)(j))$, $\Gamma_\ell(H^i_{et}(\bar{U},\mathbb{Q}_\ell(j)))$ and $\Gamma_H(H^i_B(U,\mathbb{Q}(j)))$, respectively, which is a highly non-trivial question. In fact, for $i = 2j$ this property for Γ_ℓ and Γ_H is equivalent to the conjectures of Tate and of Hodge, respectively: consider the exact sequence

$$(5.24.1)\quad H^{2j}_Y(X)(j) \xrightarrow{\mu_*} H^{2j}(X)(j) \xrightarrow{\nu^*} H^{2j}(U)(j)$$

for $\mu : Y \hookrightarrow X$ closed of codimension j and $U = X\smallsetminus Y \overset{\nu}{\hookrightarrow} X$, for the considered cohomology theory. By purity, there is a canonical isomorphism $H^{2j}_Y(X)(j) \cong \bigoplus_{y\in Y^{(0)}} 1$ (where 1 is the trivial object: $\mathbb{Q}_\ell$ with trivial G_k-action in the ℓ-adic case, the trivial Hodge structure $\mathbb{Q}$ for the Betti cohomology). This shows that 5.24.1 induces an exact sequence ($\Gamma = \Gamma_\ell$ or Γ_H, respectively)

$$(5.24.2)\qquad \Gamma(H^{2j}_Y(X)(j)) \xrightarrow{\Gamma\mu_*} \Gamma(H^{2j}(X)(j)) \xrightarrow{\Gamma\nu_*} \Gamma(H^{2j}(U)(j)) .$$

Hence, if $\Gamma\nu^* = 0$, then $\Gamma(H^{2j}(X)(j))$ is generated by cycles with support on Y.

For general i and j the situation is more complicated, since

$$(5.24.3)\qquad \Gamma(H^i_Y(X)(j)) \xrightarrow{\Gamma\mu_*} \Gamma(H^i(X)(j)) \xrightarrow{\Gamma\nu^*} \Gamma(H^i(U)(j))$$

is not necessarily exact. Nevertheless we get the following rough picture where we write $H^i_Y(X,j)$ for $H^i_Y(X)(j)$. Assume for a moment that Y is smooth, of codimension $i-j$, then we have an isomorphism $H^i_Y(X,j) \cong H^{i-2(i-j)}(Y,j-(i-j)) = H^{2j-i}(Y,2j-i)$ and a commutative diagram

$$\begin{array}{ccc} \Gamma H^{2j-i}(Y,2j-i) & \xrightarrow{\Gamma\mu_!} & \Gamma H^{i}(X,j) \\ {\scriptstyle ch_{2j-i,2j-i,Y}}\uparrow & & \uparrow{\scriptstyle ch_{i,j,X}} \\ K_{2j-1}(Y)^{(2j-i)} & \xrightarrow{\mu_!} & K_{2j-i}(X)^{(j)} \end{array}$$

with the usual Gysin morphism $\mu_!$ in the cohomology and a certain Gysin morphism $\mu_!$ for the motivic cohomology (whose construction involves the Riemann-Roch theorem, cf. 7.1 below). By 5.23 the surjectivity of $ch_{i,j,X}$ reduces to the surjectivity of $\Gamma\mu_!$ and of $ch_{2j-i,2j-i,Y}$. By 5.22 b), $K_{2j-i}(Y)^{(2j-i)}$ is strongly related to Milnor K-theory, in any case we can construct some elements in this K-group by using symbols and elements in $K_1(-)^{(1)}$ $= \mathcal{O}(-)^{\times} \otimes \mathbb{Q}$. The generic surjectivity of

$$ch_{m,m,Y} : K_m(Y)^{(m)} \longrightarrow \Gamma H^m(Y,m)$$

is related to the theorem of Merkurjev-Suslin [MS 1] saying that for any field F and integer n, char(F) $\nmid$ n, the Galois symbol

$$K_m^{Milnor}(F)/n \longrightarrow H^m_{ét}(F,\mathbb{Z}/n(m))$$

is an isomorphism for $m \leq 2$, and to the conjecture of Kato that this should be true for all $m \geq 0$. We can incorporate all this in the following drawing

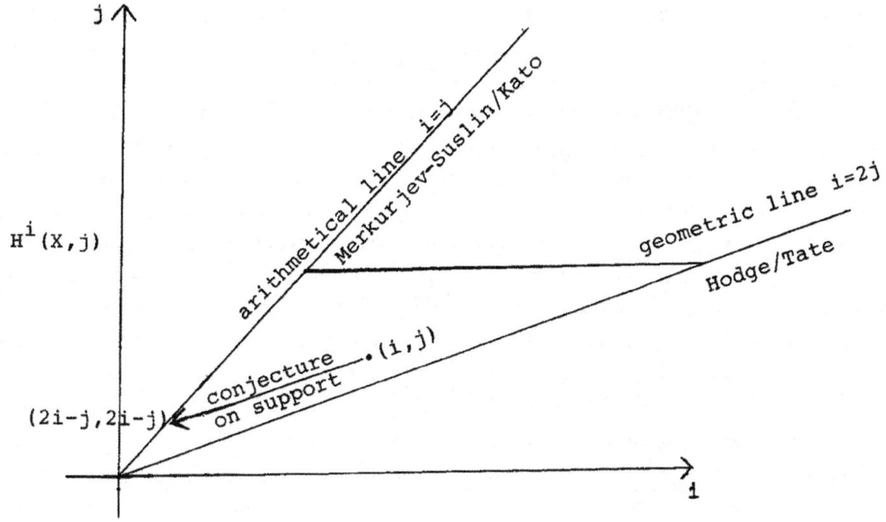

where the triangle is the area with $\Gamma H^i(X,j) \neq 0$ (possibly).

Of course, this picture is not really true as we remarked above. First of all, the vanishing of $\Gamma\nu^*$ does not imply the surjectivity of $\Gamma\mu_*$. Secondly, the subvariety will in general be singular, and we cannot argue by Gysin morphisms. Hence it turns out to be useful to study singular varieties as well, and also the non-exactness of $\Gamma = \mathrm{Hom}(1,-)$, i.e., the derivatives $R^p\Gamma = \mathrm{Ext}^p(1,-)$ of Γ for $p \geq 0$. This will be discussed in the next chapters.

§6. Twisted Poincaré duality theories

A suitable setting for our purposes is the notion of a "twisted Poincaré duality theory" as introduced by Bloch and Ogus [BO] 1.3. We need a version with values in a tensor category, not just in abelian groups.

6.1. Definition Let V be a category of schemes of finite type over a field k containing all quasi-projective ones, and let T

be an abelian tensor category in the sense of [DMOS] II 1.15, with identity object $\underline{1}$.

1) A twisted Poincaré duality theory on $\mathcal{V}$ with values in $\mathcal{T}$ is given by a collection of objects of $\mathcal{T}$

$H^i_Y(X,j)$ (cohomology with support in Y)

$H_a(X,b)$ (homology)

for every object X of $\mathcal{V}$ and every closed immersion $Y \hookrightarrow X$ in $\mathcal{V}$ and every $i,j,a,b \in \mathbb{Z}$ such that

a) $H^i_Y(X,j)$ is contravariant with respect to cartesian squares

$$\begin{array}{ccc} Y & \hookrightarrow & X \\ \downarrow & & \downarrow \\ Y' & \hookrightarrow & X' \end{array}$$

in $\mathcal{V}$ (see [BO] for a more precise description of this property and the following ones; we concentrate here rather on the necessary modifications for working with $\mathcal{T}$),

b) $H_a(X,b)$ is contravariant with respect to étale morphisms and covariant with respect to proper morphisms in $\mathcal{V}$,

c) for $Z \subseteq Y \subseteq X$ there is a long exact sequence

$$\ldots \to H^i_Z(X,j) \to H^i_Y(X,j) \to H^i_{Y \setminus Z}(X \setminus Z,j) \to H^{i+1}_Z(X,j) \to \ldots ,$$

functorial with respect to the contravariance in a),

d) (excision) for $Z \subset X$ closed and $U \subset X$ open with $Z \subseteq U$ the morphism $H^i_Z(X,j) \to H^i_Z(U,j)$ is an isomorphism,

e) if the diagram below on the left is cartesian, with proper f,g

and étale α, β , then the diagram on the right commutes

$$\begin{array}{ccc} X' & \xrightarrow{\beta} & X \\ g \downarrow & & \downarrow f \\ Y' & \xrightarrow{\alpha} & Y \end{array} \qquad\qquad \begin{array}{ccc} H_i(X,n) & \xrightarrow{\beta^*} & H_i(X',n) \\ f_* \downarrow & & \downarrow g_* \\ H_i(Y,n) & \xrightarrow{\alpha^*} & H_i(Y',n) \end{array} ,$$

f) if $Y \overset{i}{\hookrightarrow} X$ is a closed immersion and $\alpha : X \smallsetminus Y \hookrightarrow X$ is the corresponding open immersion, then there is a long exact sequence

$$\ldots \to H_a(Y,b) \xrightarrow{i_*} H_a(X,b) \xrightarrow{\alpha^*} H_a(X\ Y,b) \longrightarrow H_{a-1}(Y,b) \longrightarrow \ldots ,$$

functorial with respect to proper morphisms,

g) there is a morphism (cap-product) for $Y \hookrightarrow X$ closed

$$H_i(X,m) \otimes H^j_Y(X,n) \xrightarrow{\cap} H_{i-j}(Y,m-n) ,$$

compatible with the contravariance for étale morphisms,

h) (projection formula) for a cartesian diagram on the left with proper of the diagram on the right is commutative

$$\begin{array}{ccc} Y' & \hookrightarrow & X' \\ f \downarrow & & \downarrow f \\ Y & \hookrightarrow & X \end{array} \qquad \begin{array}{ccccc} H_i(X',m) & \otimes & H^j_{Y'}(X',n) & \longrightarrow & H_{i-j}(Y',m-n) \\ f_* \downarrow & & \uparrow f^* & & \downarrow f_* \\ H_i(X,m) & \otimes & H^j_Y(X,n) & \longrightarrow & H_{i-j}(Y,m-n) \end{array} ,$$

i) (fundamental class) for each variety X in V , which is irreducible of dimension d , there is a canonical morphism

$$\eta_X \in \mathrm{Hom}_T(\underline{1}, H_{2d}(X,d)) =: \Gamma(H_{2d}(X,d)) ,$$

which is functorial with respect to étale morphisms,

j) (Poincaré duality) if $X \in ob(V)$ is irreducible, smooth of dimension d and $Y \hookrightarrow X$ is a closed immersion, the morphism

$$H_Y^{2d-i}(X,d-n) \xrightarrow{\eta_X \cap} H_i(Y,n)$$

given by

$$H_Y^{2d-i}(X,d-n) \cong 1 \otimes H_Y^{2d-i}(X,d-n) \xrightarrow{\eta_X \otimes id} H_{2d}(X,d) \otimes H_Y^{2d-i}(X,d-n) \overset{\cap}{\to} H_i(Y,n)$$

is an isomorphism,

k) in the situation of j), for $Z \subseteq Y$ closed the diagram

$$\begin{array}{ccccccccc}
\ldots \to H_{Y \setminus Z}^{2d-i-1}(X \setminus Z,d-j) & \to & H_Z^{2d-i}(X,d-j) & \to & H_Y^{2d-j}(X,d-j) & \to & H_{Y \setminus Z}^{2d-i}(X \setminus Z,d-j) \to \ldots \\
\wr \downarrow \eta_{X \setminus Z} \cap & & \wr \downarrow \eta_X \cap & & \wr \downarrow \eta_X \cap & & \wr \downarrow \eta_{X \setminus Z} \cap \\
\ldots \to H_{i+1}(Y \setminus Z,j) & \longrightarrow & H_i(Z,j) & \longrightarrow & H_i(Y,j) & \longrightarrow & H_i(Y \setminus Z,j) \to \ldots
\end{array}$$

is commutative (this is not postulated in [BO], but will be needed below).

2) A morphism of twisted Poincaré duality theories is a pair of morphism of functors which is compatible with the axioms a)-k) in the obvious sense.

By definition, we let $H^i(X,j) = H_X^i(X,j)$.

6.2. Remark Since the definition of tensor categories is quite abstract, we like to remind the reader of the following.

a) In the cases we are interested in, the category T is usually a category of "vector spaces with some additional structure" and the tensor law is given by the tensor product of vector spaces.

b) If T is an abelian category with tensor product, i.e., where for each two objects A,B the functor

$$C \rightsquigarrow \mathrm{Bil}(A,B,C) = \{\text{bilinear morphisms } f : A \oplus B \longrightarrow C\}$$

is representable by an object $A \otimes B$:

$$\mathrm{Hom}(A \otimes B, C) = \mathrm{Bil}(A,B,C) ,$$

then $(A,B) \rightsquigarrow A \otimes B$ with the obvious commutativity and associativity constraints is a tensor law with constraints AC [SR] I 2.1.1, so it only needs an identity object $\underline{1}$ to obtain a tensor category.

<u>6.3. Definition</u> Let F be a field. An F-linear, rigid abelian tensor category $\mathcal{T}$ has a weight filtration, if there is a sequence W_m of exact subfunctors of $id : \mathcal{T} \to \mathcal{T}$ for $m \in \mathbb{Z}$ such that

a) $W_m \subset W_{m+1}$, and for every object A in $\mathcal{T}$ the filtration $W_m A$ is finite, exhausting, and separated, i.e., $W_m A = 0$ for $m << 0$ and $W_m A = A$ for $m >> 0$,

b) for objects A,B of $\mathcal{T}$ one has

$$W_m(A \otimes B) = \sum_{p+q=m} W_p A \otimes W_q B$$

(note that the sum is finite by a)).

Letting $Gr_m^W A = W_m A / W_{m-1} A$, say that the weight $m \in \mathbb{Z}$ occurs in A , if $Gr_m^W A \neq 0$, and that A is pure of weight m , if m is the only weight occuring in A .

<u>6.4. Lemma</u> The following properties follow from the axioms.

i) $\mathrm{Hom}(A,B) = 0$ if the weights occuring in A and B are distinct, e.g., if A and B are pure of different weights,

ii) $\underline{1}$ is pure of weight 0,

iii) $Gr_m^W(A \otimes B) \cong \bigoplus_{p+q=n} Gr_p^W A \otimes Gr_q^W B$,

iv) $W_{-m}(A^\vee) \cong (A/W_{m-1}A)^\vee$ (where $B^\vee$ is the dual of B ([DMOS]II 1.6)).

Proof i) By induction on the exact sequences

$$0 \longrightarrow W_{m-1}C \longrightarrow W_mC \longrightarrow Gr_m^W C \longrightarrow 0$$

for $C = A$ and $C = B$ it suffices to consider the case that A and B are pure of weights $m \neq n$, say. Since the functors Gr_m^W are exact, too, we get for a morphism $f : A \longrightarrow B$ that $\mathrm{Ker}\, f \cong Gr_m^W \mathrm{Ker}\, f \cong Gr_m^W A \cong A$, hence $f = 0$.

ii) By 6.3 a) and decomposing $\underline{1}$ and $\mathcal{T}$ if necessary (cf. [DMOS] II.1.17) we may suppose that $\underline{1}$ is pure of weight m , say. By 6.3 b) and the isomorphism $\underline{1} \otimes \underline{1} \simeq \underline{1}$ we conclude $m = 0$.

iii) This follows from 6.3 b) and the exactness of the tensor product.

iv) This follows from the exactness of $A \leadsto A^\vee$ and the fact that $A^\vee$ is pure of weight $-m$, if A is pure of weight m : in this case we have $W_{-m-1}A^\vee = 0$, since $\mathrm{Hom}(W_{-m-1}A^\vee, A^\vee) = \mathrm{Hom}(W_{-m-1}A^\vee \otimes A, 1) = 0$, and $X := A^\vee/W_{-m}A^\vee = 0$, since $X^\vee \subseteq A^{\vee\vee} = A$ is pure of weight m and hence $\mathrm{Hom}(X^\vee, X^\vee) = \mathrm{Hom}(X^\vee \otimes X, 1) = 0$. q.e.d.

For the following let R be a commutative ring (with unit) and let $\mathcal{T}$ be an abelian, R-linear tensor category. If F is the field of fractions of R , we obtain a new abelian, F-linear tensor category $\mathcal{T} \otimes F$, which has the same objects as $\mathcal{T}$, and where the morphism sets are defined by

$$\mathrm{Hom}_{\mathcal{T}\otimes F}(A_F, B_F) = \mathrm{Hom}_{\mathcal{T}}(A,B) \otimes_R F \,,$$

where A_F is the object of $\mathcal{T} \otimes F$ associated to $A \in \mathrm{ob}(\mathcal{T})$. Say that $\mathcal{T}$ has a weight filtration if $\mathcal{T} \otimes F$ has.

6.5. Definition A twisted Poincaré duality theory with values in $\mathcal{T}$ has weights, if $\mathcal{T} \otimes F$ is rigid and has a weight filtration, and if the following conditions hold.

a) The weights w occuring in $H_a(X,b)_F$ satisfy

$$2b-a \leq w \leq 2b \qquad \text{for } a \leq d = \dim X \,,$$

$$2b-a \leq w \leq 2b-2(a-d) \qquad \text{for } a \geq d \,.$$

b) For X/k proper the weights w occuring in $H^i(X,j)_F$ satisfy

$$-2j \leq w \leq i-2j \qquad \text{for } i \leq d = \dim X \,,$$

$$2(i-d)-2j \leq w \leq i-2j \qquad \text{for } i \leq d \,.$$

6.6. Corollary For X/k smooth the weights w occuring in $H^i(X,j)_F$ satisfy

$$i-2j \leq w \leq 2i-2j \qquad \text{for } i \leq d = \dim X \,,$$

$$i-2j \leq w \leq 2d-2j \qquad \text{for } i \leq d \,.$$

In particular, for X smooth and proper over k, $H^i(X,j)_F$ is pure of weight $i-2j$.

This is clear from 6.5 and the Poincaré duality isomorphisms 6.1 j). We now give some examples.

<u>6.7. Example</u> Let $\mathcal{V}$ be the category of all separated schemes of finite type over a field k with separabel closure k_s, let $G_k = \mathrm{Gal}(k_s/k)$ be its absolute Galois group, and let ℓ be a prime different from $\mathrm{char}(k)$. The category $\mathrm{Rep}_c(G_k,\mathbb{Z}_\ell)$ of finitely generated $\mathbb{Z}_\ell$-modules with continuous action of G_k is an abelian, $\mathbb{Z}_\ell$-linear tensor category: the tensor law is the tensor product over $\mathbb{Z}_\ell$, and the identity object is $\mathbb{Z}_\ell$ with trivial operation; note that we have

$$\Gamma(M) = \mathrm{Hom}_{G_k}(\mathbb{Z}_\ell,M) \cong M^{G_k} \quad \text{by} \quad \varphi \longmapsto \varphi(1) .$$

We get a twisted Poincaré duality on $\mathcal{V}$ with values in $\mathrm{Rep}_c(G_k,\mathbb{Z}_\ell)$ by letting

$$\begin{aligned} H^i_Z(X,j) &= H^i_{\text{ét},\bar{Z}}(\bar{X},\mathbb{Z}_\ell(j)) \quad \text{for} \quad Z \hookrightarrow X \quad \text{closed} , \\ H_a(X,b) &= H_a^{\text{ét}}(\bar{X},\mathbb{Z}_\ell(b)) \\ &= H^{-a}_{\text{ét}}(\bar{X},R\bar{f}^!\mathbb{Z}_\ell(-b)) \quad \text{for} \quad X \xrightarrow{f} \mathrm{Spec}\, k \end{aligned} \tag{6.7.1}$$

(ℓ-adic étale cohomology and homology, cf. [BO] 2.1 and [DV] exp. VIII), where $\bar{X} = X \times_k \bar{k} \xrightarrow{\bar{f}} \mathrm{Spec}\,\bar{k}$ denotes the base extension to the algebraic closure $\bar{k}$ of k. The category $\mathrm{Rep}_c(G_k,\mathbb{Z}_\ell)$ is equivalent to the category of constructible $\mathbb{Z}_\ell$-sheaves on $\mathrm{Spec}\, k$, and the finite generation of the above $\mathbb{Z}_\ell$-modules follows from Deligne's result that in the above situation Rf_* and $Rf^!$ respect (complexes of) constructible sheaves, see [SGA $4\frac{1}{2}$] [finitude] 2.9. $\mathrm{Rep}_c(G_k,\mathbb{Z}_\ell) \otimes \mathbb{Q}_\ell$ can be identified with the category $\mathrm{Rep}_c(G_k,\mathbb{Q}_\ell)$ of finite-dimensional $\mathbb{Q}_\ell$-vector spaces with continuous action of G_k (equivalent to the category of constructible $\mathbb{Q}_\ell$-sheaves on $\mathrm{Spec}\, k$) and is rigid: the internal Hom is given by

$$\underline{\mathrm{Hom}}(V,W) = \mathrm{Hom}_{\mathbb{Q}_\ell}(V,W)$$

with the G_k-action $(\sigma f)(u) = \sigma f(\sigma^{-1}v)$.

6.8. Example Now let k be finitely generated. Say that $V \in \mathrm{ob}(\mathrm{Rep}_c(G_k,\mathbb{Q}_\ell))$ has a weight filtration, if there exists an integral domain A of finite type over $\mathbb{Z}[\frac{1}{\ell}]$ with field of fractions k such that V extends to a constructible $\mathbb{Q}_\ell$-sheaf F over $U = \mathrm{Spec}\, A$, which has an increasing exhausting and separating filtration W_mF by constructible subsheaves such that $Gr_m^W F = W_mF/W_{m-1}F$ is pointwise pure of weight m , see [D9] 1.2.2. Since F is smooth ("constant tordue") on a neighbourhood of the generic point $\eta = \mathrm{Spec}\, k$ of U , this amounts to saying that

i) there is a connected, smooth scheme U' over $\mathbb{Z}[\frac{1}{\ell}]$ (char $k = 0$) or $\mathbb{F}_p$ (char $k = p > 0$) such that the representation of G_k on V factorizes through $G_k \twoheadrightarrow \pi_1(U',\bar{\eta})$, $\bar{\eta} = \mathrm{Spec}\, \bar{k}$,

ii) there is an increasing exhausting and separating filtration $\ldots \subseteq W_{m-1}V \subseteq W_mV \subseteq \ldots$ of G_k-submodules such that for each closed point $x \in U'$ the eigenvalues of a geometric Frobenius Fr_x at x in $\pi_1(U,\bar{\eta})$ on $Gr_m^W V = W_mV/W_{m-1}V$ are algebraic numbers α with absolute value

$$|\alpha| = Nx^{\frac{m}{2}} \ , \ Nx = \#\kappa(x) \ ,$$

for every archimedean valuation $||$. Here the geometric Frobenius in $\mathrm{Gal}(\overline{\kappa(x)}/\kappa(x))$ is the inverse of the arithmetic Frobenius $a \longmapsto a^{Nx}$, and its image in $\pi_1(U',\bar{\eta})$ via $\mathrm{Gal}(\overline{\kappa(x)}/\kappa(x)) = \pi_1(\kappa(x),\overline{\kappa(x)}) \longrightarrow \pi_1(U,\overline{\kappa(x)}) \cong \pi_1(U,\bar{\eta})$ is well-defined up to conjugacy. For a finite field k we have $U' = \mathrm{Spec}\, k$, and for a global field k with ring of integers $\mathcal{O}_k$ we have $U' = \mathrm{Spec}\, \mathcal{O}_k \setminus S$ for a finite set of primes S including all primes above ℓ . Then

$\pi_1(U',\bar{\eta})$ is the Galois group G_S of the maximal S-ramified extension of k, and i) means that V is unramified outside S.

A filtration as in ii) is called weight filtration for V.

6.8.1. Lemma a) A weight filtration, if it exists, is unique.
b) Let $WRep_c(G_k,\mathbb{Q}_\ell)$ be the full subcategory of $Rep_c(G_k,\mathbb{Q}_\ell)$ formed by the representations having a weight filtration. Then every morphism on $WRep_c(G_k,\mathbb{Q}_\ell)$ is strictly compatible ([D4] 1.1.5) with the weight filtrations, and $WRep_c(G_k,\mathbb{Q}_\ell)$ is an abelian subcategory of $Rep_c(G_k,\mathbb{Q}_\ell)$, closed with respect to taking subobjects or quotients.
c) $WRep_c(G_k,\mathbb{Q}_\ell)$ is a rigid, abelian, $\mathbb{Q}_\ell$-linear tensor category with weights.

Proof It follows immediately from the definitions that there is no non-trivial G_k-morphism between pure modules of different weights, hence the same statement for distinct sets of weights. From this one deduces that every morphism is compatible with the given weight filtrations and the unicity (look at the identity map). Subobjects and quotients obtain a weight filtration by the induced and the quotient filtration, respectively (cf. [D4] 1.1.8), hence $WRep_c(G_k,\mathbb{Q}_\ell)$ is an abelian subcategory of $Rep_c(G_k,\mathbb{Q}_\ell)$. Since every isomorphism is an isomorphism of filtered objects by the above, we obtain the strictness for every morphism. This in turn implies that the functors $V \rightsquigarrow W_mV$ are exact (cf. [D4] 1.1.11 and the remark after it). The claims in c) are now clear: the filtration on $\underline{Hom}(V,W)$ is given by

$$W_m\underline{Hom}(V,W) = \{f : V \to W \mid f(W_iV) \subseteq W_{i+m}W \text{ for all } i\},$$

and the filtration on $V \otimes W$ is given by the formula in 6.3 b).

<u>6.8.2. Lemma</u> Denote by $WRep_c(G_k,\mathbb{Z}_\ell)$ the full subcategory of $Rep_c(G_k,\mathbb{Z}_\ell)$ formed by those objects M for which $M \otimes_{\mathbb{Z}_\ell} \mathbb{Q}_\ell$ has a weigth filtration. Then the functors 6.7.1 have image in $WRep_c(G_k,\mathbb{Z}_\ell)$ and form a twisted Poincaré duality theory with weights.

<u>Proof</u> For a closed immersion $Z \hookrightarrow X$ of algebraic k-schemes there is a smooth scheme U_0 over $\mathbb{Z}[\frac{1}{\ell}]$ (if $\operatorname{char} k = 0$) or over $\mathbb{F}_p$ (if $\operatorname{char} k = p > 0$), with generic point $\eta = \operatorname{Spec} k$, and a closed dimension

$$\begin{array}{ccc} Z & \overset{\nu}{\hookrightarrow} & X \\ & {}_{h}\searrow \qquad \swarrow_{f} & \\ & U_0 & \end{array}$$

of separated U_0-schemes of finite type such that $Z \hookrightarrow X$ is obtained from ν by base change to $\operatorname{Spec} k$.

By Deligne's generic base change theorem (cf. [SGA $4\frac{1}{2}$] [finitude] 1.5 and 2.9) the operations Rf_*, $Rf_!$, $Rf^!$, $R\nu^!$ and $R\nu_*$ respect constructible complexes, and are compatible with arbitrary base change $S \longrightarrow U$ for a suitable open subscheme U of U_0 . In particular, $H^i_{\bar{Z}}(\bar{X},\mathbb{Z}_\ell(j))$ extends to the constructible sheaf $F = H^i(Rh_*R\nu^!\mathbb{Z}_\ell(j))$, and $H_a(\bar{X},\mathbb{Z}_\ell(b))$ extends to the constructible sheaf $G = H^{-a}(Rf_*Rf^!\mathbb{Z}_\ell(-b))$ on U . Moreover, the associated $\mathbb{Q}_\ell$-sheaves are mixed by Deligne's result [D9] 6.1.11.

This shows the claim for $\operatorname{char}(k) = p > 0$, since by loc.cit. 3.4.1 every mixed sheaf in this case also has a weight filtration. The bounds on the weights follow from loc.cit. 3.3.8, since by the base change property we have

$$H_x = H^a_c(X \times_{U_0} \overline{\kappa(x)}\, ,\, \mathbb{Z}_\ell(b))$$

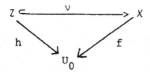

(cohomology with compact support) for $\mathcal{H} = R^a f_! \mathbb{Z}_\ell$ and every point x of U, and since

(6.8.3) $$H_a^{\text{ét}}(\bar{X}, \mathbb{Q}_\ell(b)) \cong H^a_{\text{ét},c}(\bar{X}, \mathbb{Q}_\ell(b))^\vee$$

($\mathbb{Q}_\ell$-dual) by the duality $\underline{RHom}(Rf_! \mathbb{Q}_\ell, \mathbb{Q}_\ell) \cong Rf_* \underline{RHom}(\mathbb{Q}_\ell, Rf^! \mathbb{Q}_\ell) \cong Rf_* Rf^! \mathbb{Q}_\ell$.

For char $k = 0$ we cannot argue in this way, since the concepts of mixed sheaves and weight filtrations are different here (see the remark below). Instead we shall get the weight filtration by resolution of singularities, and this will be proved in 6.11 together with the result for absolute Hodge cycles (note that we may assume all sheaves as being reduced by the topological invariance of étale cohomology).

<u>6.8.4. Remarks</u> i) Let k be a number field, let S be a finite set of primes including all primes above ℓ, and let $G_S = \pi_1(\text{Spec } \mathcal{O}_k \smallsetminus S, \bar{\eta})$ as above (ramification at infinity is allowed). Then for $n \in \mathbb{Z}$ one easily computes

$$\dim_{\mathbb{Q}_\ell} H^1_{\text{cont}}(G_S, \mathbb{Q}_\ell(2n+1)) \geq r_1 + r_2 \geq 1$$

where r_1 and r_2 are the numbers of real and complex places of k, respectively (cf. [J3] lemma 2), so by the isomorphism $H^1_{\text{cont}}(G_S, \mathbb{Q}_\ell(m)) \cong \text{Ext}^1_{G_S}(\mathbb{Q}_\ell, \mathbb{Q}_\ell(m))$ there exist non-trivial extensions of continuous G_S-representations

(6.8.5) $$0 \longrightarrow \mathbb{Q}_\ell(2n+1) \longrightarrow E \longrightarrow \mathbb{Q}_\ell \longrightarrow 0 .$$

Since $\mathbb{Q}_\ell$ and $\mathbb{Q}_\ell(2n+1)$ are pure of weights 0 and $-2(2n+1)$, respectively, E corresponds to a mixed sheaf on Spec $\mathcal{O}_k \setminus S$. For $n < 0$, however, E cannot have a weight filtration since $W_0 E$ would give a splitting of 6.8.5.

ii) Let E_S be the group of S-units and Cl_S be the S-class group of k , then there are exact sequences

$$0 \longrightarrow E_S/\ell^n \longrightarrow H^1(G_S, \mu_{\ell^n}) \longrightarrow {}_{\ell^n}Cl_S \longrightarrow 0 ,$$

for all $n \geq 1$. Since by Dirichlet's theorem E_S is a finitely generated group and Cl_S is finite, we conclude

$$H^1_{cont}(G_S, \mathbb{Z}_\ell(1)) \cong E_S \otimes \mathbb{Z}_\ell ,$$

by passing to the limit over n . On the other hand we have

$$H^1_{cont}(G_k, \mathbb{Z}_\ell(1)) \cong \hat{k}^\times = \lim_{\underset{n}{\longleftarrow}} k^\times/k^{\times^{\ell^n}} .$$

Since $\hat{k}^\times \supsetneqq k^\times \otimes \mathbb{Z}_\ell = \lim_{\underset{S}{\longrightarrow}} E_S \otimes \mathbb{Z}_\ell$, we conclude that

$$(6.8.6) \qquad \lim_{\underset{S}{\longrightarrow}} H^1_{cont}(G_S, \mathbb{Q}_\ell(1)) \subsetneqq H^1_{cont}(G_k, \mathbb{Q}_\ell(1)) .$$

This shows that there is a (non-trivial) extension of continuous G_k-representations

$$(6.8.7) \qquad 0 \longrightarrow \mathbb{Q}_\ell(1) \longrightarrow E' \longrightarrow \mathbb{Q}_\ell \longrightarrow 0 ,$$

which does not come from a G_S-extension for any S . Hence E' corresponds to a $\mathbb{Q}_\ell$-sheaf on Spec k which does not come from Spec$(\mathcal{O}_k \setminus S)$ for any S , this answers the question in [D9] 6.1.1. The example is the same as Serre's example in [Se 1] III 2.2.

6.9. Example Let $A = \mathbb{Z}$, $\mathbb{Q}$ or $\mathbb{R}$, then the category $A\text{-}\mathcal{MH}$ of mixed A-Hodge structures is an A-linear abelian tensor category with weights; the identity object is the trivial Hodge strcuture A (pure of weight zero), and we have

$$\Gamma(H) = \mathrm{Hom}_{A\text{-}\mathcal{MH}}(A,H) \cong W_0H \cap F^0H_{\mathbb{C}} =: \Gamma_{\mathcal{H}}(H)$$

$$f \longmapsto f(1) .$$

For the category $\mathcal{V}$ of all separated schemes of finite type over $\mathbb{C}$ the Betti cohomology and homology

$$H^i_Z(X,j) = H^i_{Z(\mathbb{C})}(X(\mathbb{C}),A)(j)$$

(6.9.1)

$$H_a(X,b) = H_a(X(\mathbb{C}),A)(-b) \text{ (Borel-Moore homology)}$$

with the mixed A-Hodge structure associated to it by Deligne forms a twisted Poincaré duality theory with weights. Note that $H^i_Z(X,A)$ has a mixed Hodge structure as relative cohomology $H^i(X,X\smallsetminus Z,A)$ ([D5] 8.3.8), for the Borel-Moore (or Betti) homology we may use the isomorphism

(6.9.2) $$H_a(X(\mathbb{C}),\mathbb{Q}) \cong H^a_c(X(\mathbb{C}),\mathbb{Q})^\vee$$

($\mathbb{Q}$-dual of the mixed $\mathbb{Q}$-Hodge structure given by the cohomology with compact support) and the relation $H^a_c(X,A) = H^a(\tilde{X},Z,A)$ for a compactification $X \subseteq \tilde{X}$ with closed complement $Z = \tilde{X}\smallsetminus X$; see also [J2] § 2 for a direct approach. The sign (m) denotes the Tate twist in Hodge theory [D4] 2.1.13, and the compatibility of the mixed Hodge structures with exact sequences and products follows as in [D5] 8.1.25 and 8.3.9. The bounds on the weights for a proper scheme are proved in [D5] 8.2.4, the bounds for homology follow via 6.7.9

from corresponding bounds for cohomology with compact support, which are mentioned in [D7] 8 and follow from the fact that the cohomology in question is the cohomology of a simplicial scheme with smooth and proper components, and that $H^i_c(U,\mathbb{Z}) = 0$ for a smooth affine variety of pure dimension $d > i$.

<u>6.10. Example</u> Let k be a field of characteristic zero, and let $\mathcal{V}$ be the category of varieties (i.e., reduced separated schemes of finite type) over k. Then

$$\begin{aligned} H^i_Z(X,j) &= H^i_{DR,Z}(X/k) \\ H_a(X,b) &= H^{DR}_a(X/k) \end{aligned} \qquad (6.10.1)$$

(de Rham cohomology with support and de Rham homology) gives a twisted Poincaré duality theory with values in the category Vec_k of finite dimensional vector spaces over k. If X is <u>smooth</u> one may set

$$H^i_{DR,Z}(X,k) = H^i_Z(X,\Omega^{\cdot}_{X/k}) \quad \text{(Zariski hypercohomology)}$$

and if X is embeddable in a smooth scheme M of pure dimension N (e.g., if X is quasi-projective), one may set

$$H^{DR}_a(X/k) = H^{2N-a}_{DR,X}(M/k) .$$

In general one has to use resolution of singularities and calculate the above groups for suitable simplicial schemes with smooth components (see 6.11 below, but also the approach in [Harts]).

<u>6.11. Example</u> The following generalizes the constructions for

absolute Hodge cycles carried out in the first part. Let k be a finitely generated field of characteristic zero, then we define the category IMR_k of integral mixed realizations for absolute Hodge cycles over k like the category $\underline{MR}_k$ in § 2, except for the following changes: We replace the filtered $\mathbb{Q}_\ell$-representations of G_k by objects of $W\,\mathrm{Rep}_c(G_k,\mathbb{Z}_\ell)$, requiring that the filtration $W.$ of 2.1 b) is in fact a weight filtration in the sense of example b) above. Furthermore, we replace mixed $\mathbb{Q}$-Hodge structures by mixed Hodge structures and postulate that the comparison isomorphisms of 2.1 e) are integrally defined, i.e., as isomorphisms $H_\sigma \otimes_{\mathbb{Z}} \mathbb{Z}_\ell \xrightarrow{\sim} H_\ell$ of finitely generated $\mathbb{Z}_\ell$-modules. Finally, there should exist morphisms $Gr^W_m H_{\mathbb{Q}} \otimes Gr^W_m H_{\mathbb{Q}} \longrightarrow \underline{1}_{\mathbb{Q}}(-m)$ for all $m \in \mathbb{Z}$ defining polarizations of the real Hodge structures $Gr^W_m H_\sigma \otimes \mathbb{R}$.

It is clear that IMR_k is a $\mathbb{Z}$-linear tensor category with weight filtration, and the following result generalizes 3.19.

<u>6.11.1. Theorem</u> Let $\mathcal{V}$ be the category of varieties over k. Then there is a twisted Poincaré duality theory with weights and values in IMR_k on $\mathcal{V}$

$$H^*_{AH,Z}(X,*),\ H^{AH}_*(X,*)$$

such that the de Rham components H_{DR}, the ℓ-adic components H_ℓ and the Betti components H_σ for $\sigma : k \hookrightarrow \mathbb{C}$ are given by the functors in 6.10.1, 6.7.1 and 6.9.1 for $\sigma X = X \times_{k,\sigma} \mathbb{C}$, respectively, and such that for a smooth quasi-projective variety U over k $H^n_{AH}(U,0)$ coincides with the mixed realization $H^n(U)$ defined in 3.19 after tensoring with $\mathbb{Q}$.

<u>Proof</u> We define functors $H^i_{AH,Z}(X)$ and $H^{AH}_a(X)$ and then let

$$H^i_{AH,Z}(X,j) = H^i_{AH,Z}(X)(j) ,$$

$$H^{AH}_a(X,b) = H^{AH}_a(X)(-b) .$$

The ℓ-adic and the Betti realizations of $H^i_{AH,Z}(X)$ and $H^{AH}_a(X)$ are

(6.11.2)
$$H^i_{\text{ét},\bar{Z}}(\bar{X},\mathbb{Z}_\ell) \quad \text{and} \quad H^{\text{ét}}_a(\bar{X},\mathbb{Z}_\ell) ,$$
$$H^i_{\sigma Z}(\sigma X,\mathbb{Z}) \quad \text{and} \quad H^{BM}_a(\sigma X,\mathbb{Z}) ,$$

respectively, then the claimed compatibilities with 6.7 and 6.9 are clear. The comparison isomorphisms between them are defined as in § 3, by the canonical comparison isomorphisms

(6.11.3)
$$H^i_{\sigma Z}(\sigma X,\mathbb{Z}) \otimes \mathbb{Z}_\ell \xrightarrow{\sim} H^i_{\text{ét},\sigma Z}(\sigma X,\mathbb{Z}_\ell) ,$$
$$H^{BM}_a(\sigma X,\mathbb{Z}) \otimes \mathbb{Z}_\ell \xrightarrow{\sim} H^{\text{ét}}_a(\sigma X,\mathbb{Z}_\ell) ,$$

which in fact are integrally defined and also exist for relative cohomology and homology as indicated (cf. [SGA4] XVI 4.1 and [DV] VI 2.8.4). For defining the de Rham realizations, the other comparison isomorphisms, and the weight filtrations compatible with the comparison isomorphisms we use "simplicial resolutions" as in [D5].

First it suffices and is more general to define homology for simplicial varieties with proper face and degeneration maps and cohomology for arbitrary simplicial varieties, since homology of varieties (as the homology of the associated constant simplicial variety) and cohomology with support (as relative cohomology for a morphism $f. : X. \longrightarrow U.$ and hence as cohomology of the simplicial variety $\mathrm{Cone}(f)$, see [D5] 6.3) are special cases of this.

For a simplicial variety $Z.$ as above we define the ℓ-adic and the Betti realizations of $H^i_{AH}(Z.)$ and $H^{AH}_a(Z.)$ by the simplicial versions of 6.11.2, and the comparison isomorphisms by the simplicial versions of 6.11.3 (for their existence compare the arguments below). For a morphism $\pi : U. \longrightarrow Z.$ of two such simplicial varieties this is functorial in a contravariant way for cohomology, and functorial in a covariant way for homology if π is proper.

Now there exists a proper morphism $\pi : U. \longrightarrow Z.$ inducing an isomorphism in the étale cohomology and homology (e.g., by proper hypercoverings, cf. [D5] 6.2 and 8.3) such that $U.$ is the complement in a smooth and proper simplicial variety $X.$ of a divisor $Y.$ with normal crossings. Basically, everthing reduces to a simplicial version of § 3. Namley, by the arguments in [D5] it suffices to define H^i_{AH} and H^{AH}_a functorially for $U.$, since by the comparison isomorphisms a morphism will induce an isomorphism of all groups, if it induces an isomorphism in étale cohomology and homology. In fact, it suffices to define H^i_{AH} and H^{AH}_a functorially for pairs $(X.,Y.)$ as above such that the Betti realizations are those of $U.$, cf. Beilinson's constructions in [Bei 1] § 1.

Thus we define

$$\begin{aligned} H^i_{DR}(Z.) &:= H^i_{Zar}(U.,\Omega^{\cdot}_{U./k}) = H^i_{Zar}(X.,\Omega^{\cdot}_{X.}\langle Y.\rangle) , \\ (6.11.4) \qquad & \\ H^{DR}_a(Z.) &:= H^a_{DR,c}(Z.)^{\vee} \quad \text{(k-dual)} \end{aligned}$$

where the de Rham cohomology with compact support is defined as relative cohomology

$$(6.11.5) \qquad H^a_{DR,c}(Z.) := H^a_{DR,c}(U.) := H^a_{DR}(X.,Y.)$$

by the remark above; one gets quite explicit complexes computing these groups by a similar approach as in [J2] 2.9. By the comparison isomorphisms in the smooth case, generalized to the complexes computing $H^i_{an}(\sigma U.,\Omega^{\cdot an}_{\sigma U./k})$ and $H^i(\sigma U.,\mathbb{Z})$, we obtain the comparison isomorphisms

$$(6.11.6)\qquad I_{\infty,\sigma} : H^i(\sigma U.,\mathbb{Z}) \otimes_{\mathbb{Z}} \mathbb{C} \longrightarrow H^i_{Zar}(U.,\Omega^{\cdot}_{U./k}) \otimes_{k,\sigma} \mathbb{C} .$$

If we define the weight and the Hodge filtrations by the formulae in [D5] 8.1.12 and 8.1.15, we see that they are respected by the comparison isomorphisms - by formula [D5] 7.1.6.5 and the results in § 3 for $W.$, and by the comparison isomorphisms

$$H^i_{Zar}(X.,F^r\Omega^{\cdot}_{X./k}<Y.>)\otimes_{k,\sigma}\mathbb{C}\cong H^i_{Zar}(\sigma X.,F^r\Omega^{\cdot}_{\sigma X./\mathbb{C}}<\sigma Y.>)\cong H^i_{an}(\sigma X.,F^r\Omega^{\cdot}_{\sigma X.}<\sigma Y.>^{an})$$

for $F^{\cdot}$. If we also use the mentioned formulae to put a weight filtration on $H^i_{ét}(\bar{U}.,\mathbb{Z}_\ell)$, we get a G_k-equivariant ascending filtration, compatible with the comparison isomorphisms

$$(6.11.7)\qquad I_{\ell,\bar{\sigma}} : H^i(\sigma U.,\mathbb{Z}) \otimes_{\mathbb{Z}} \mathbb{Z}_\ell \longrightarrow H^i_{ét}(\bar{U}.,\mathbb{Z}_\ell) .$$

It is a weight filtration in the sense of b) above, since the (i+n)-th graded term is a subquotient of

$$\bigoplus_m H^{i-n-2m}_{ét}(\overline{Y_m^{(n+m)}}, \mathbb{Q}_\ell(n+m))$$

after tensoring with $\mathbb{Q}_\ell$, cf. [D5] 8.1.19 b).

For obtaining the long exact sequences of definition 6.1 and the capproduct it is convenient to proceed like Beilinson for the Hodge theory in [Bei 2]. Namely, by the same arguments as in loc.cit. § 4 one in fact gets much more than just the cohomology groups above and the comparison isomorphisms between them. By Deligne's constructions in [D5] 8.1, now also applied to the étale and the algebraic de Rham cohomology, one gets for each object $(X.,Y.)$ as above a diagram

(6.11.8) $\quad K^{\bullet} = K^{\bullet}_{AH}(X.,Y.)$:

$$\begin{array}{ccccc} & \prod\limits_{\ell}\ \prod\limits_{\sigma:k\hookrightarrow C} ({}'K^{\bullet}_{\ell,\sigma},W) & & \prod\limits_{\sigma:k\hookrightarrow \mathbb{C}} ({}'K^{\bullet}_{\mathbb{C},\sigma},W) & \\ \alpha = \prod\limits_{\ell,\sigma}\alpha_{\ell,\sigma} \nearrow & & \nwarrow \beta = \prod\limits_{\ell,\sigma}\beta_{\ell,\sigma} \quad \gamma = \prod\limits_{\sigma}\gamma_{\sigma} \nearrow & & \nwarrow \delta = \prod\limits_{\sigma}\delta_{\sigma} \\ \prod\limits_{\ell}(K^{\bullet}_{\ell},W) & & \prod\limits_{\sigma:k\hookrightarrow \mathbb{C}}(K^{\bullet}_{\sigma},W) & & (K^{\bullet}_{k},W,F)\ , \end{array}$$

called a <u>mixed absolute Hodge complex</u> (MAH-complex), of the following kind:

i) For each ℓ, $K^{\bullet}_{\ell}$ is a bounded below, filtered complex of $\mathbb{Z}_{\ell}$-modules with continuous G_k-action such that the homology groups are finitely generated $\mathbb{Z}_{\ell}$-modules.

ii) ${}'K^{\bullet}_{\ell,\sigma}$, for each ℓ and each $\sigma : k \hookrightarrow \mathbb{C}$, is as in i), but without action of G_k,

iii) $K^{\bullet}_{\sigma}$ (resp. ${}'K^{\bullet}_{\mathbb{C},\sigma}$), for each $\sigma : k \hookrightarrow \mathbb{C}$, is as in ii), but with $\mathbb{Z}_{\ell}$ replaced by $\mathbb{Z}$ (resp. $\mathbb{C}$),

iv) $K^{\bullet}_{k}$ is a bounded below complex of k-vector spaces, with finite dimensional homology, an ascending filtration W and a descending filtration F,

v) for each $\sigma : k \hookrightarrow \mathbb{C}$,

$$\alpha_{\sigma,\ell} = (\alpha_{\ell,\bar{\sigma}})_{\bar{\sigma}:k \hookrightarrow \mathbb{C},\ \bar{\sigma}|_k = \sigma}$$

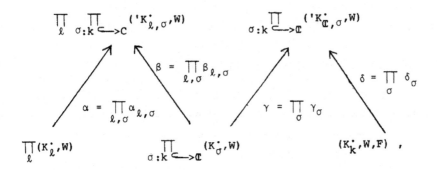

is a family of filtered quasi-isomorphisms

$$\alpha_{\ell,\bar{\sigma}} : (K^{\bullet}_{\ell}, W) \longrightarrow ({}'K^{\bullet}_{\ell,\sigma}, W)$$

with $\alpha_{\ell,\bar{\sigma}\rho}\rho \simeq \alpha_{\ell,\bar{\sigma}}$ (homotopic) for $\rho \in G_k$,

vi) for each (σ,ℓ) ,

$$\beta_{\ell,\sigma} : (K^{\bullet}_{\sigma}, W) \otimes_{\mathbb{Z}} \mathbb{Z}_{\ell} \longrightarrow ({}'K^{\bullet}_{\ell,\sigma}, W)$$

is a filtered quasi-isomorphism,

vii) for each $\sigma : k \hookrightarrow \mathbb{C}$,

$$\gamma_{\sigma} : (K^{\bullet}_{\sigma}, W) \otimes_{\mathbb{Z}} \mathbb{C} \longrightarrow ({}'K^{\bullet}_{\mathbb{C},\sigma}, W)$$

is a filtered quasi-isomorphism,

viii) for each $\sigma : k \hookrightarrow \mathbb{C}$,

$$\delta_{\sigma} : (K^{\bullet}_{k}, W) \otimes_{k,\sigma} \mathbb{C} \longrightarrow ({}'K^{\bullet}_{\mathbb{C},\sigma}, W)$$

is a filtered quasi-isomorphism,

ix) $H^i(Gr^W_m K^{\bullet}_{\ell})$ is pure of weight $m+i$, and for each $\sigma : k \hookrightarrow \mathbb{C}$

$$\begin{array}{ccc} & ({}'K^{\bullet}_{\mathbb{C},\sigma}, W) & \\ {}^{\gamma_{\sigma}}\nearrow & & \nwarrow^{\delta_{\sigma}} \\ (K^{\bullet}_{\sigma}, W) & & (K^{\bullet}_{k}, W, F) \otimes_{k,\sigma} \mathbb{C} \end{array}$$

defines a mixed Hodge complex in the sense of [D5] 8.1.5.

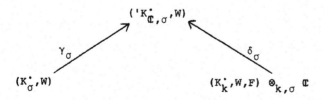

The topology referred to in i) is the ℓ-adic one; it is actually better to consider pro-ℓ-systems ([J1] 6.9) of discrete G_k-modules instead - and pro-ℓ-systems of abelian groups in ii) instead of $\mathbb{Z}_\ell$-modules. The comparison quasi-isomorphism in vi) then must be interpreted in an appropriate way, by letting $(K_\sigma^\cdot, W) \otimes_{\mathbb{Z}} \mathbb{Z}_\ell$ be a canonical filtered complex of pro-ℓ-systems whose components (in the pro-ℓ-direction) are quasi-isomorphic to $K_\sigma^\cdot \otimes^L \mathbb{Z}/\ell^n$.

Deligne has proved that the i-th cohomology of a mixed Hodge complex with the filtrations induced by F and $W[i]$ is a mixed Hodge structure. From this it is obvious that the i-th cohomology of a mixed absolute Hodge complex defines an object in $\mathcal{MR}_k$. We define polarized MAH-complexes by requiring that the graded pieces $H = Gr_m^W H^i \otimes \mathbb{Q}$ for the weight filtration of these objects are polarizable in the sense that we have bilinear forms $H_? \otimes H_? \longrightarrow \mathbb{Q}_?(-m)$ in each realization which are compatible under the comparison isomorphisms and define polarizations of the real Hodge structures H_σ . There is an obvious notion of morphisms between MAH-complexes, of tensor products, a trivial MAH-complex $\mathbb{Z}$, and Tate twists. Following Beilinson's constructions for polarized mixed Hodge complexes (called $\tilde{p}$-Hodge complexes by him in [Bei 2] § 3), one has a notion of cones, homotopy, and quasi-isomorphisms for polarized MAH-complexes, and thus obtains a triangulated category of MAH-complexes up to quasi-isomorphism.

For a simplicial object $(X.,Y.)$ as above - $X.$ a smooth and proper simplicial variety and $Y. \subseteq X.$ a divisor with normal crossings - and $U. = X.\smallsetminus Y.$ one obtains the polarized MAH-complex $K_{AH}^\cdot(X.,Y.)$ as the diagram

(6.11.9)

$$\begin{array}{ccccc}
 & \prod\limits_{\ell}\prod\limits_{\sigma}(R\Gamma_{et}(\sigma U_\cdot,\mathbb{Z}_\ell),W) & & \prod\limits_{\sigma}(R\Gamma_{an}(\sigma U_\cdot,\Omega^{\cdot an}_{\sigma U_\cdot}),W) & \\
\nearrow & & \nwarrow \quad \nearrow & & \nwarrow \\
\prod\limits_{\ell}(R\Gamma_{et}(\bar{U}_\cdot,\mathbb{Z}_\ell),W) & & \prod\limits_{\sigma}(R\Gamma_{an}(\sigma U_\cdot,\mathbb{Z}),W) & & (R\Gamma_{Zar}(X_\cdot,\Omega^\cdot_X\langle Y_\cdot\rangle),W,F)\ ,
\end{array}$$

where W and F are defined by the procedure of [D5] 8.1, and the maps come from the morphisms $\sigma U_\cdot \xrightarrow{\bar{\sigma}} \bar{U}_\cdot$, $\mathbb{Z} \longrightarrow \Omega^{\cdot an}_{\sigma U_\cdot}$, and the comparison isomorphisms, respectively, the latter ones extending to simplicial schemes by the standard computation of their cohomology ([D5] 5.2.7). The careful reader will notice that the comparison isomorphisms in 6.11.9 are only naturally defined in the derived category, since they involve quasi-isomorphisms in the wrong direction. To define them as honest morphisms of complexes one may proceed as Beilinson in [Bei 2] p. 55, by replacing the upper complexes by canonical quasi-isomorphic ones.

By working on the big sites one then obtains the complexes $K^\cdot_{AH}(X_\cdot,Y_\cdot)$ functorial (contravariant) in $(X_\cdot,Y_\cdot)$. Starting from this, one may copy Beilinson's arguments in [Bei 2] § 4 to get polarized MAH-complexes whose cohomology gives the previously defined H^i_{AH} and H^{AH}_a of varieties, of simplicial varieties, or versions $H^i_{AH,c}$ with compact support, with the required properties. For example, $H^i_{AH}(Z_\cdot)$ is computed by $K^\cdot_{AH}(X_\cdot,Y_\cdot)$ for $X_\cdot \smallsetminus Y_\cdot = U_\cdot \xrightarrow{\pi} Z_\cdot$ as above (for a more sophisticated version giving canonical complexes $K^\cdot_{AH}(Z_\cdot)$ one may proceed as in [Bei 1] 1.6.5). $H^i_{AH,c}(Z_\cdot)$ is computed by $\mathrm{Cone}(K^\cdot_{AH}(X_\cdot) \longrightarrow K^\cdot_{AH}(\Delta\tilde{Y}_{\cdot\cdot}))[-1]$, where $\tilde{Y}_{\cdot\cdot}$ is the coskeleton of the normalization $\tilde{Y}_\cdot \longrightarrow Y$. (cf. [J2] 2.9) and $\Delta\tilde{Y}_{\cdot\cdot}$ is the diagonal. $H^{AH}_a(Z_\cdot)$ is computed by

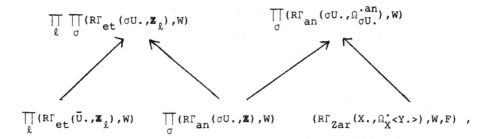

(6.11.10) $\quad K^{AH}_\cdot(Z_\cdot) = R\,\underline{\mathrm{Hom}}(K^\cdot_{AH,c}(Z_\cdot),\mathbb{Z})$, where $K^{\cdot}_{AH,c}(Z.)$ is a complex computing $H^{i}_{AH,c}(Z.)$ - e.g., the previous one, and R <u>Hom</u> is the obvious derived internal Hom (replace the complex $\mathbb{Z}$ by one with injective components and take internal Hom for each single complex).

The long exact sequences required in 6.1 c) and f), including the restriction map for open immersions in homology now come from carefully chosen simplicial resolutions and exact triangles of MAH-complexes by the same arguments as in [Bei 1] 1.8, cf. also [J2] 1.19. This assures that the maps in the exact cohomology sequences are morphisms in $\mathcal{IMR}_k$. For defining a capproduct it suffices by 6.11.10 to define a pairing

$$K^{\cdot}_{AH,Z}(X) \otimes^{L}_{\mathbb{Z}} K^{\cdot}_{AH,c}(Z) \longrightarrow K^{\cdot}_{AH,c}(X) \ . \tag{6.11.11}$$

By the mentioned triangles and the following lemma, applied to

$$\begin{array}{ccccc} K^{\cdot}_{AH}(X) & \otimes^{L}_{\mathbb{Z}} & K^{\cdot}_{AH,c}(Z) & \xrightarrow{\ \psi\ } & K^{\cdot}_{AH,c}(X) \\ \downarrow & & \uparrow & & \uparrow \\ K^{\cdot}_{AH}(X \smallsetminus Z) & \otimes^{L}_{\mathbb{Z}} & K_{AH,c}(X \smallsetminus Z) & \xrightarrow{\ \psi'\ } & K_{AH,c}(X \smallsetminus Z) \end{array} \ , \tag{6.11.12}$$

we see that it suffices to define a pairing ψ as indicated functorial in the sense of 6.11.12 for an arbitrary variety X . By (carefully chosen) simplicial resolutions one reduces to the case that X is smooth, and then one may proceed as Beilinson in [Bei 2] 4.2; everything carries over to MAH-complexes.

<u>6.11.13. Lemma</u> Given compatible pairings of complexes (of abelian groups, G_k-modules, ...), ψ and ψ' , as indicated, there is a canonical pairing φ making the following diagram commutative

$$\begin{array}{ccccc}
{}'A[-1] & \otimes & {}'B^{\cdot}[1] & \xrightarrow{\ \tilde{\psi}'\ } & {}'C^{\cdot} \\
\downarrow & & \uparrow & & \downarrow \gamma \\
\mathrm{Cone}(\alpha)[-1] & \otimes & \mathrm{Cone}(\beta) & \xrightarrow{\ \varphi\ } & C^{\cdot} \\
\downarrow & & \uparrow & & \| \\
A & \otimes & B & \xrightarrow{\ \psi\ } & C^{\cdot} \\
\alpha \downarrow & & \uparrow \beta & & | \gamma \\
A' & \otimes & B' & \xrightarrow{\ \psi'\ } & {}'C^{\cdot} \ .
\end{array}$$

Here $\tilde{\psi}'$ is the canonical pairing induced by ψ' : on ${}'A^{p-1} \otimes {}'B^{q+1}$ it is $(-1)^{p-1}\psi'$.

<u>Proof</u> Let $\varphi((a',a) \otimes (b.b')) = \psi(a \otimes b) + (-1^{p-1}\gamma\ \psi'(a' \otimes b')$ on $\mathrm{Cone}(\alpha)[-1]^p \otimes \mathrm{Cone}(\beta)^q = ({}'A^{p-1} \oplus A^p) \otimes (B^q \oplus {}'B^{q+1})$.

With these definition it is straightforward to check that $H^i_{AH,Z}(X)(j)$ and $H^{AH}_a(X)(-b)$ form a twisted Poincaré duality theory, except that we have not defined the restriction map for ètale morphisms in homology. Since this restriction map is only needed for open immersions in the sequel, and since its construction is somewhat involved, we don't give it here. Let us just remark that the compatibility of [BO] 1.4.2 holds (this can be checked in any of the realizations and is known to hold in the étale theory),hence all results of Bloch and Ogus in loc.cit. apply.

Similarly, the Poincaré duality morphism, defined by the capproduct with the fundamental class, is an isomorphism since it is an isomorphism in the ℓ-adic realization. The fundamental class can be obtained as in [BO] 2.1, by reducing to the smooth case and then taking the preimage of $1 \in \Gamma(H^0_{AH}(X,0)) = \mathbb{Z}$ under

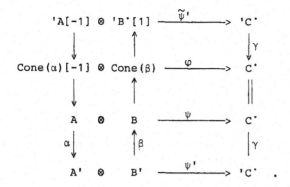

$$\begin{array}{ccc} H_{2\dim X}(X, \dim X) \otimes H^0(X,0) & \xrightarrow[\sim]{\cap} & H^0(X,0) \\ \| & & \\ H_{2\dim X}(X, \dim X) \ . & & \end{array}$$

All other properties required in 6.1 easily follow from the case of the ℓ-adic realizations, via the comparison isomorphisms.

<u>6.12. Example</u> It is indicated in [Bei 1] and proved in [Sou 3] théorème 9, that on the category $\mathcal{V}_S$ of schemes quasi-projective over a regular noetherian irreducible base S (it should also be universally caténaire) the <u>motivic cohomology and homology</u>

$$H^i_{M,Z}(X,\mathbb{Q}(j)) = K_{2j-1}(X)^{(j)} ,$$

(6.12.1)

$$H^M_a(X,\mathbb{Q}(b)) = K'_{a-2b}(X)^{(-b)}$$

form a twisted Poincaré duality theory, with the restriction that $H^i_{M,Z}(X,\mathbb{Q}(j))$ with $Z \underset{\neq}{\subset} X$ is perhaps only defined for smooth X (In the contrast to the previous examples, this is an "absolute" theory, without weights). Here K' denotes Quillen's K-theory for coherent sheaves, and $K'_m(X)^{(r)}$ is defined as the subspace in $K'_m(X) \otimes \mathbb{Q}$, where the Adams operators $'\psi^k$ act by multiplication with k^r for all k (cf. the definition of $K_m(X)^{(r)}$ in 5.12). For the definition of the $'\psi^k$ (called ϕ^k in [Sou 3]) we refer the reader to the cited papers, it involves the embedding of X into a scheme W smooth over S and then changing the Adams operators on $K^X_m(W) \xleftarrow{\sim} K'_m(X)$ by multiplication with cannibalistic classes of $\Omega^1_{W|S}$ to make everything independent of W.

Actually, Soulé in his paper defines motivic cohomology and homology by means of the descending γ-filtration on K_m and an ascending γ-filtration on $K'_m \otimes \mathbb{Q}$, setting

$$H^i_{M,Z}(X,\mathbb{Q}(j)) = Gr^j_\gamma K^Z_{2j-i}(X) \otimes \mathbb{Q} ,$$

(6.12.2)

$$H^M_a(X,\mathbb{Q}(b)) = Gr^\gamma_b(K'_{a-2b}(X) \otimes \mathbb{Q}) .$$

Both approaches are equivalent, by the canonical isomorphisms

$$K^Z_m(X)^{(r)} \xrightarrow{\sim} Gr^r_\gamma K_m(X) \otimes \mathbb{Q} ,$$

(6.12.3)

$$K'_m(X)^{(r)} \xrightarrow{\sim} Gr^\gamma_{-r}(K'_m(X) \otimes \mathbb{Q}) ,$$

induced by the inclusions $K^Z_m(X)^{(r)} \subseteq \gamma^r K_m(X) \otimes \mathbb{Q}$ and $K'_m(X)^{(r)} \subseteq \gamma_{-r} K'_m(X) \otimes \mathbb{Q}$, cf. [Sou 3] 2.8 and 7.2. However, one has to be careful with using this identification, cf. the discussion in [J2] 3.7.

In lower degrees - for the K-groups - one has the following calculation of motivic (co-)homology.

<u>6.12.4. Lemma</u> Quillen's niveau spectral sequence ([Q1] § 7, 5.4)

(6.12.5) $$E^1_{p,q}(X) = \bigoplus_{x \in X_{(p)}} K_{p+q}(\kappa(x)) \Rightarrow K'_{p+q}(X) ,$$

where $X_{(p)}$ is the set or points of X with dimension p and $\kappa(x)$ is the residue field of $x \in X$, induces isomorphisms

a) $H^M_{2b}(X,\mathbb{Q}(b)) \cong E^2_{b,-b}(X) \otimes \mathbb{Q} = CH_b(X) \otimes \mathbb{Q}$, where $CH_b(X)$ is the Chow "homology" group of cycles of dimension b , as defined by Fulton [Fu],

b) $H^M_{2b+1}(X,\mathbb{Q}(b)) \cong E^2_{b+1,-b}(X) \otimes \mathbb{Q}$, with $E^2_{b+1,-b}(X)$ = homology of

$$\bigoplus_{x \in X_{(b+2)}} K_2(\kappa(x)) \xrightarrow{\text{tame}} \bigoplus_{x \in X_{(b+1)}} \kappa(x)^\times \xrightarrow{\text{div}} \bigoplus_{x \in X_{(b)}} \mathbb{Z} ,$$

where "tame" denotes the tame symbol and "div" the divisor map,

c) $H^{\mathcal{M}}_{2d-2}(X,\mathbb{Q}(d-2)) \cong E^2_{d,2-d}(X) \otimes \mathbb{Q}$ for $d = \dim X$, with

$$E^2_{d,2-d}(X) = \mathrm{Ker}\big(\bigoplus_{x\in X_{(d)}} K_2(\kappa(x)) \xrightarrow{\text{tame}} \bigoplus_{x\in X_{(d-1)}} \kappa(x)^{\times}\big) .$$

In particular, for X irreducible, smooth of dimension d over a field k the Brown-Gersten spectral sequence

(6.12.6) $$E_2^{p,q}(X) = H^p_{Zar}(X,\mathcal{K}_{-q}) \Rightarrow K_{-p-q}(X) ,$$

where $\mathcal{K}_m$ is the Zariski sheaf associated to $U \longmapsto K_m(U)$, induces isomorphisms

d) $H^{2j}_{\mathcal{M}}(X,\mathbb{Q}(j)) \cong H^j(X,\mathcal{K}_j) \otimes \mathbb{Q} \cong CH^j(X) \otimes \mathbb{Q}$,

e) $H^{2j-1}_{\mathcal{M}}(X,\mathbb{Q}(j)) \cong H^{j-1}(X,\mathcal{K}_j) \otimes \mathbb{Q} \cong \mathbb{Q} \otimes$ homology of

$$\bigoplus_{x\in X^{(j-2)}} K_2(\kappa(x)) \xrightarrow{\text{tame}} \bigoplus_{x\in X^{(j-1)}} \kappa(x)^{\times} \xrightarrow{\text{div}} \bigoplus_{x\in X^{(j)}} \mathbb{Z} ,$$

f) $H^2_{\mathcal{M}}(X,\mathbb{Q}(2)) \cong H^0(X,\mathcal{K}_2) \otimes \mathbb{Q} \cong \mathbb{Q} \otimes$ kernel of

$$K_2(k(X)) \xrightarrow{\text{tame}} \bigoplus_{x\in X^{(1)}} \kappa(x)^{\times} ,$$

where $k(X)$ is the function field of X.

<u>Proof</u> The first three statements follow from the spectral sequence constructed from 6.12.5 by Soulé in [Sou 3] théorème 8 iv), in view of the fact that

$$K_0(F) = Gr^0_{\gamma}K_0(F) = \mathbb{Z} ,$$

(6.12.7) $$K_1(F) = Gr^1_{\gamma}K_1(F) = F^{\times} ,$$

$$K_2(F) = Gr^2_{\gamma}K_2(F) = K_2^{Milnor}(F) \text{ (Milnor K-theory)}$$

for a field F (cf. also loc.cit. théorème 4 and remarque), and since $d^1_{p+1,-p}$ = div and $d^2_{p+2,-p}$ = tame in 6.12.5 (cf. [Q1] § 7, 5.6). For the same reason one has

$$CH_b(X) = \mathrm{Coker}\Big(\bigoplus_{x \in X_{(b+1)}} \kappa(x)^\times \xrightarrow{\mathrm{div}} \bigoplus_{x \in X_{(b)}} \mathbb{Z}\Big) = E^2_{b,-b}(X) .$$

In view of the construction of 6.12.6 from 6.12.5 by means of the Gersten resolution (see [Q1] § 7, 5.8), i.e., by the formulae

$$E_r^{p,q}(X) = E^r_{d-p,d-q}(X), \quad X^{(p)} = X_{(d-p)} ,$$

the statements d), e) and f) are just a reformulation.

For irreducible X the fundamental class $\eta_X^{\mathcal{M}} \in H^{\mathcal{M}}_{2d}(X,\mathbb{Q}(d))$, $d = \dim X$, is defined as the element corresponding to the class of X in $CH_d(X)$ via 6.12.4 a). In Soulé's description this is the class of $\mathcal{O}_X$ in $Gr^{\gamma}_d(K'_0(X) \otimes \mathbb{Q})$, however, in general the class in $K'_0(X)^{(-d)}$ corresponding to this will be different.

§7. The conjectures of Hodge and Tate for singular varieties

For a twisted Poincaré duality theory with values in a tensor category $\mathcal{T}$ the fundamental classes induce a cycle map

(7.1.1) $$\begin{aligned} cl_i : Z_i(X) &\longrightarrow \Gamma H_{2i}(X,i) \\ [Z] &\longmapsto \text{image of } \eta_Z \text{ under} \\ &\qquad \Gamma H_{2i}(Z,i) \longrightarrow \Gamma H_{2i}(X,i) \end{aligned}$$

from the group of cycles of dimension i into the "group of sections" of the homology group $H_{2i}(X,i)$. Here

$$Z_i(X) = \bigoplus_{x \in X_{(i)}} \mathbb{Z}$$

is identified with the free abelian group on the irreducible closed subvarieties $Z \subset X$ of dimension i, via $x \mapsto Z = \{\overline{x}\}$. For X smooth, irreducible of dimension d 7.1.1 by Poincaré duality (cf. 6.1 j))

$$H_{2i}(X,i) \cong H^{2d-2i}(X,d-i)$$

reads as

$$(7.2.2) \qquad cl^j : Z^j(X) \longrightarrow \Gamma H^{2j}(X,j)$$

where $j = d - i$ is now the codimension of the cycles. As explained in §5, the conjectures of Hodge and Tate concern the image of the last map.

I claim that for an arbitrary variety X the correct generalization consists in stating similar conjectures for the maps 7.1.1. Note that for singular varieties there is not even a reasonable cycle map into cohomology, since the morphism $H^{2d-2i}(X,d-i) \xrightarrow{\eta_X \cap} H_{2i}(X,i)$ is not an isomorphism in general.

7.2. Conjecture (Hodge conjecture for singular varieties) If X is a variety (i.e., a separated, reduced algebraic scheme) over $\mathbb{C}$, then for all $i \geq 0$ the map

$$cl_i \otimes \mathbb{Q} : Z_i(X) \otimes \mathbb{Q} \longrightarrow \Gamma_*(H_{2i}(X,\mathbb{Q})(i)))$$

$$= (2\pi\sqrt{-1})^{-i} W_{-2i} H_{2i}(X,\mathbb{Q}) \cap F^{-i} H_{2i}(X,\mathbb{C})$$

is surjective (the homology being the Borel-Moore one).

7.3. Conjecture (Tate conjecture for singular varieties) If k is a finitely generated field and X is a variety over k, then for each prime $\ell \neq \mathrm{char}\,(k)$ and every $i \geq 0$ the map

$$cl_i \otimes \mathbb{Q}_\ell : Z_i(X) \otimes \mathbb{Q}_\ell \longrightarrow H_{2i}^{\acute{e}t}(\overline{X},\mathbb{Q}_\ell(i))^{G_k}$$

is surjective (where $H_a^{\acute{e}t}(\overline{X},\mathbb{Q}_\ell(b)) = H_a^{\acute{e}t}(\overline{X},\mathbb{Z}_\ell(b)) \otimes_{\mathbb{Z}_\ell} \mathbb{Q}_\ell$, the notations being those of 6.7).

We want to show that 7.2 and 7.3 basically reduce to the classical conjectures for smooth projective varieties. For this we more generally study the cycle map 7.1.1 for a twisted Poincaré duality theory with weights on a category $\mathcal{V}$ of schemes over a field k, as defined in 6.5.

7.4. Remark If for our Poincaré duality theory in addition we assume (cf. [BO] 1.5):

ℓ) (principal triviality) If $i : D \longrightarrow X$ is a smooth principal divisor in a smooth variety X, then $i_*\eta_D = 0$,

which is satisfied in the examples of §6, then the cycle map factorizes through rational equivalence as defined by Fulton, inducing a map

$$cl_i : CH_i(X) \longrightarrow \Gamma H_{2i}(X,i),$$

cf. [BO] p. 197, step 1.

7.5. Lemma Let $U \xrightarrow{\alpha} X$ be an open immersion. Then the restriction

$$W_{2b-a}H_a(X,b)_F \xrightarrow{\alpha^*} W_{2b-a}H_a(U,b)_F$$

is surjective.

Proof. Let $Z \subset X$ be a closed subscheme with $U = X - Z$, then by 6.1 f) and the exactness of W we have an exact sequence

$$W_{2b-a}H_a(X,b)_F \xrightarrow{\alpha^*} W_{2b-a}H_a(U,b)_F \longrightarrow 0,$$

since $W_{2b-a}H_a(Z,b)_F = 0$ by 6.5 a).

In the following we assume the following property for our Poincaré duality theory (compare [BO] 7.1.2)

m) If $f : X \longrightarrow Y$ is a proper map between varieties of the same dimension, then $f_*\eta_X = (\deg f)\ \eta_Y$.

Also we assume that $\mathcal{T}$ is an F-linear category for the field $F = H^0$ (Spec k,0), which is of characteristic zero. All this is satisfied for the Hodge theory $(F = \mathbb{Q})$ and the ℓ-adic theory $(F = \mathbb{Q}_\ell)$.

7.6. Lemma Assume that k is perfect and that all schemes in $\mathcal{V}$ are reduced. If $\pi : X' \longrightarrow X$ is proper and surjective, then

$$W_{2b-a}H_a(X',b) \xrightarrow{\pi_*} W_{2b-a}H_a(X,b)$$

is surjective for all $a,b \in \mathbb{Z}$.

<u>Proof.</u> There exists a smooth subvariety $U \subsetneq X$, open and dense in every irreducible component, such that the restriction of $\pi : U' = \pi^{-1}(U) \longrightarrow U$ is faithfully flat. By treating the connected components separately we may assume that U is connected. Then there is a commutative diagram

$$\begin{array}{ccc} & & U' \\ & \overset{g}{\nearrow} & \\ U'' & & \downarrow \pi \\ & \underset{h}{\searrow} & \\ & & U \end{array}$$

with h faithfully flat, quasi-finite separated (cf. [Mi] I 2.25). By Zariski's Main Theorem and possibly making U (and U') smaller we may assume that h is flat and finite. Then, by assumption m),

$$\eta_U = (\deg h)^{-1}\, h_*\eta_{U''} = (\deg h)^{-1}\pi_*(g_*\eta_{U''})$$

(note that g is necessarily proper), i.e., η_U is in the image of π_*. By the commutative diagram

$$\begin{array}{ccc} H^i(U',j) & \xrightarrow{(\deg h)^{-1}g_*\eta_{U''}\cap} & H_{2d-i}(U',d-j) \\ \pi^* \uparrow & & \downarrow \pi_* \\ H^i(U,j) & \underset{\sim}{\xrightarrow{\eta_U\cap}} & H_{2d-i}(U,d-j) \end{array}, \quad d = \dim U,$$

we see that π_* has a right inverse, so the claim is true for $\pi : U' \longrightarrow U$.

For X we proceed by induction on the dimension. The case $\dim X = 0$ is covered by the considerations above. Otherwise let U be as above, $Z = X \setminus U$ and $Z' = \pi^{-1}(Z) = X' \setminus U'$. Then by 6.1 f) and 7.5 we obtain a commutative exact diagram

$$\begin{array}{ccccccc} W_{2b-a}H_a(Z',b) & \longrightarrow & W_{2b-a}H_a(X',b) & \longrightarrow & W_{2b-a}H_a(U',b) & \longrightarrow & 0 \\ \pi_* \downarrow & & \pi_* \downarrow & & \pi_* \downarrow & & \\ W_{2b-a}H_a(Z,b) & \longrightarrow & W_{2b-a}H_a(X',b) & \longrightarrow & W_{2b-a}H_a(U',b) & \longrightarrow & 0 \ . \end{array}$$

The left π_* is surjective by assumption of the induction, the right π_* is so by the first step, hence the surjectivity in the middle.

<u>7.7. Remark</u> Let $X' \xrightarrow{\pi} X$ be proper and surjective with X proper and X' smooth and proper. Then by the above

$$\mathrm{Im}\ (H_a(X',b) \xrightarrow{\pi_*} H_0(X,b)) = W_{2b-a}H_a(X,b).$$

In particular, in the situations of 6.8 and 6.9, where $H_a(-,b)$ is the dual of the cohomology with compact suport $H_c^a(-,b)$, we obtain

$$\mathrm{Ker}(H^i_{\acute{e}t}(\bar{X},\mathbb{Q}_\ell) \longrightarrow H^i_{\acute{e}t}(\bar{X}',\mathbb{Q}_\ell)) = W_{i-1}H^i_{\acute{e}t}(\bar{X},\mathbb{Q}_\ell),$$

$$\mathrm{Ker}(H^i(X(\mathbb{C}),\mathbb{Q}) \longrightarrow H^i(X'(\mathbb{C}),\mathbb{Q}) = W_{i-1}H^i(X(\mathbb{C}),\mathbb{Q}),$$

respectively. The second property is [D5] 8.2.5, and the first one generalizes it to étale cohomolgy and arbitrary fields, without the assumption of resolution of singularities.

7.8. Now consider a diagram of varieties

$$(7.8.1)\qquad \begin{array}{ccc} X' & \xrightarrow{\alpha} & X'' \\ {\scriptstyle\pi}\downarrow & & \\ X & & \end{array}$$

with π proper and surjective, α an open immersion and X'' smooth and proper. Then by 7.5 and 7.6 the composition

$$(7.8.2)\qquad H_{2i}(X'',i) \xrightarrow{\alpha^*} W_0H_{2i}(X',i) \xrightarrow{\pi_*} W_0H_{2i}(X,i)$$

is surjective, and we get a commutative diagram

$$(7.8.3)\qquad \begin{array}{ccccccc} \Gamma H_{2i}(X'',i) & \xrightarrow{\Gamma\alpha^*} & \Gamma W_0H_{2i}(X',i) & \xrightarrow{\Gamma\pi_*} & \Gamma W_0H_{2i}(X,i) & = & \Gamma H_{2i}(X,i) \\ \uparrow{\scriptstyle cl_i} & & \uparrow{\scriptstyle cl_i} & & \uparrow{\scriptstyle cl_i} & & \\ Z_i(X'')\otimes F & \xrightarrow{\alpha^*} & Z_i(X')\otimes F & \xrightarrow{\pi_*} & Z_i(X)\otimes F\ . & & \end{array}$$

If $\Gamma\pi_* \circ \Gamma\alpha^*$ is still surjective, then the surjectivity of the left cycle map implies that of the right one. Since Γ is not exact, it is in general difficult to decide when this will be the case, but it is obiously true, if

a) $W_0H_{2i}(X,i)$ is (via $\pi_* \circ \alpha^*$) a direct factor of $H_{2i}(X'',i)$.

In this case we say that 7.8.1 (or X'', for short) is a good proper cover of X (for the considered homology theory H_*). In particular this holds true, if

b) $H_{2i}(X'',i)$ is a semi-simple object of $\mathcal{T}$.

7.9. Theorem The Hodge conjecture is true for arbitrary varieties, if it is true for smooth and projective ones.

Proof. By Chow's lemma and resolution of singularities, for a given $X/\mathbb{C}$ there exists a diagram 7.8.1 with smooth and projective X''. Then $H_{2i}(X''(\mathbb{C}),\mathbb{Q}(i)) \cong H^{2d-2i}(X''(\mathbb{C}),\mathbb{Q}(d-i))$, $d = \dim X''$, is a pure polarized Hodge structure, hence semi-simple. By 7.8.3 the Hodge conjecture for X'' implies the Hodge conjecture for X.

Similarly, we have:

7.10. Theorem a) For a finitely generated field k of characteristic zero the Tate conjecture 7.3 for arbitrary varieties is true if the original Tate conjecture 5.1 is true for smooth projecitve varieties.
b) For a finitely generated field k of characteristic $p \geq 0$ the Tate conjecture 7.3 is true for X, if X has a good proper cover X'' such that the Tate conjecture is true for X''.

Proof b) is clear by the above arguments, so we have to show the existence of good projective covers in the case char $k = 0$. As above, we get a diagram 7.8.1 by Chow's lemma and resolution of singularities, with X'' smooth and projective. By a general conjecture of Serre and Grothendieck, $H_{2i}^{\acute{e}t}(\overline{X''},\mathbb{Q}_\ell(i)) \cong H_{\acute{e}t}^{2d-2i}(\overline{X''},\mathbb{Q}_\ell(d-i))$, $d = \dim X''$, should be a semi-simple G_k-module, so that 7.8 b) should apply. Not knowing this, we can still get 7.8 a) by using absolute Hodge cycles in characteristic zero. Namely, as we have seen in lemma 1.1, the map 7.8.2 has a right inverse for the corresponding realizations for absolute Hodge cycles and so in _particular_ for the ℓ-adic realizations.

So far we have only discussed the first part of Tate's conjecture, which consists of the following three parts.

7.11. Conjecture (Tate [T1]) Let X be smooth, projective over the finitely generated field k and let

$$cl^j : Z^j(X) \longrightarrow H^{2j}_{ét}(\bar{X},\mathbb{Q}_\ell(j))^{G_k}$$

be the cycle map, for $\ell \neq \mathrm{char}\, k$. then

A) $cl^j \otimes \mathbb{Q}_\ell$ is surjective,

B) $A^j(X) := \mathrm{Im}\, cl^j$ is finitely generated and $A^j(X) \otimes \mathbb{Q}_\ell \xrightarrow{\sim} H^{2j}_{ét}(\bar{X},\mathbb{Q}_\ell(j))^{G_k}$,

C) $\mathrm{rk}\, A^j(X)$ = order of pole of $L(H^{2j}_{ét}(\bar{X},\mathbb{Q}_\ell),s)$ at $s = j + \dim_a k$.

_7.12._ Here $\dim_a k = \mathrm{tr.\,deg}(k)$ (+1 if char $k = 0$) is the _arithmetical_ (or Kronecker) _dimension_ of k, and "the" L-function $L(V,s)$ associated to the ℓ-adic representation $V = H^{2j}_{ét}(\bar{X},\mathbb{Q}_\ell)$ is defined as follows: there exists a smooth scheme S over $\mathbb{Z}[\frac{1}{\ell}]$ (if char $k = 0$) or $\mathbb{F}_p$ (if char $k = p > 0$) with generic point η = Spec k sucht that V extends to a pure $\mathbb{Q}_\ell$-sheaf $\mathcal{F}$ of weight $2j$ on S and we let

$$L(V,s) = \prod_{y \in |S|} \frac{1}{\det(1-Fr_y(Ny)^{-s}|\mathcal{F}_y)}, \quad s \in \mathbb{C},$$

where $|S|$ is the set of closed points of S, and, for $y \in S$, Ny is the cardinality of the finite residue field $\kappa(y)$ of y, $Fr_y \in Gal(\overline{\kappa(y)}/\kappa(y))$ is a (geometric) Frobenius, and $\mathcal{F}_y$ is the fibre of $\mathcal{F}$ at y (meaning the fibre at a geometric point $Spec\ \overline{\kappa(y)} \longrightarrow S$ over y together with the action of $Gal(\overline{\kappa(y)}/\kappa(y))$).

In fact, there is an S as above such that $X \longrightarrow Spec\ k$ extends to a smooth, projective morphism $f : \mathfrak{X} \longrightarrow S$, and then we may take $\mathcal{F} = R^{2j}f_*\mathbb{Q}_\ell$: By the proper and smooth base change theorem (cf. [Mi] VI 4.2) $\mathcal{F}$ is smooth and one has a $Gal(\overline{\kappa(y)}/\kappa(y))$-isomorphism

$$\mathcal{F}_y = H^{2j}_{ét}(\mathfrak{X} \times_S \overline{\kappa(y)}, \mathbb{Q}_\ell).$$

This is pure of weight $2j$ by Deligne's proof of the Weil conjectures [D8]. The property of the weights assures that $L(V,s)$ converges for $Re(s) > \dim S + j = \dim_a k + j$, cf. [D9] 1.4.6.

This depends on the choice of S, but the pole order at $j + \dim_a k$ doesn't, because for two choices the quotient of the L-functions is convergent for $Re(s) > \dim S - 1 + j$ (cf. Tate's argument loc. cit.).

Part B) and C) can also be generalized to arbitrary varieties, in the following form.

<u>7.13. Conjecture</u> (Tate conjectures for singular varieties). Let X be a variety over the finitely generated field k and let

$$cl_i : Z_i(X) \longrightarrow H^{ét}_{2i}(\overline{X}, \mathbb{Q}_\ell(i))^{G_k}$$

be the cycle map, for $\ell \neq \mathrm{char}\, k$. Then

A) $cl_i \otimes \mathbb{Q}_\ell$ is surjective,

B) $A_i(X) = \mathrm{Im}\, cl_i$ is finitely generated and

$A_i(X) \otimes \mathbb{Q}_\ell \xrightarrow{\sim} H_{2i}(\overline{X}, \mathbb{Q}_\ell(i))^{G_k}$,

C) $\mathrm{rk}\, A_i(X)$ = order of pole of $L(H_c^{2i}(\overline{X}, \mathbb{Q}_\ell), s)$ at $s = i + \dim_a k$.

<u>7.14.</u> $L(V,s)$ for $V = H_c^{2i}(\overline{X}, \mathbb{Q}_\ell)$ is formed as above, but now with a mixed sheaf $\mathcal{F}$ of weights $\leq 2i$ extending V over an S as before. Again this suffices to make $L(V,s)$ convergent for $\mathrm{Re}(s) > i + \dim_a k$ and the pole order independent of the choice of S. The existence of such a $\mathcal{F}$ follows as in the proof of 6.8.2, by extending $X \longrightarrow \mathrm{Spec}\, k$ to a morphism $f : \mathfrak{X} \longrightarrow U_0$ and using Deligne's generic base change theorem and his result that for a finite field H_c^a is mixed of weights $\leq a$. In fact, with the notations there we may take $S = U$ and $\mathcal{F} = R^{2i} f_! \mathbb{Z}_\ell |_U$.

For smooth and projective X, 7.11 and 7.13 are equivalent: we may assume that X is connected, of dimension d, then we have

$$A^j(X) = A_{d-j}(X), \quad H^{2j}(\overline{X}, \mathbb{Q}_\ell(j)) = H_{2d-2j}(\overline{X}, \mathbb{Q}_\ell(d-j)),$$

and furthermore

$$H_c^{2i}(\overline{X}, \mathbb{Q}_\ell) = H^{2i}(\overline{X}, \mathbb{Q}_\ell) \cong H^{2d-2i}(\overline{X}, \mathbb{Q}_\ell(d - 2i))$$

by hard Lefschetz, hence $L(H_c^{2i}(\overline{X}, \mathbb{Q}_\ell), s) = L(H^{2(d-i)}(\overline{X}, \mathbb{Q}_\ell), s + d - 2i)$. For the singular case we show:

<u>7.15. Theorem</u> Assume that X has a good proper cover X'' for $H_*^{ét}$ (e.g., let char $k = 0$). If

i) Tate B) is true for X'' and dimension i, and

ii) Tate A) is true for $X'' \times X''$ and $d'' = \dim X''$,

then Tate B) is true for X and dimension i.

<u>Proof</u> Consider the diagram

$$\begin{array}{ccccc} \epsilon : \circlearrowright & H_{2i}(\overline{X''},\mathbb{Q}_\ell(i))^{G_k} & \xrightarrow{\Gamma_\ell(\pi_*\alpha^*)} & H_{2i}(\overline{X},\mathbb{Q}_\ell(i))^{G_k} & \\ & \uparrow{\scriptstyle \varphi_{X''}} & & \uparrow{\scriptstyle \varphi_X} & \\ E : \circlearrowright & A_i(X'') & \xrightarrow{\pi_*\alpha^*} & A_i(X) & . \end{array}$$

By assumption, there is an idempotent ϵ in $\mathrm{End}_{G_k}(H_{2i}(\overline{X''},\mathbb{Q}_\ell(i)))$ with $\mathrm{Ker}\,\epsilon = \mathrm{Ker}\,\pi_*\alpha^*$. By the morphism

$$\mathrm{End}_{G_k}(H_{2i}(\overline{X''},\mathbb{Q}_\ell(i))) = \mathrm{End}_{G_k}(H^{2(d''-i)}(\overline{X''},\mathbb{Q}_\ell(d' - i)))$$

$$=[H^{2i}(\overline{X''},\mathbb{Q}_\ell) \otimes H^{2(d''-i)}(\overline{X}'',\mathbb{Q}_\ell)(d'')]^{G_k} \quad \text{(Poincaré duality)}$$

$$\subseteq H^{2d''}(\overline{X''\times X''},\mathbb{Q}_\ell(d''))^{G_k} \qquad \text{(Künneth formula)},$$

ϵ corresponds to a Tate cycle on $X'' \times X''$, so by ii) is the image of a cycle $E \in Z^{d''}(X'' \times X'')$. The algebraic correspondence E on X'' operates on $CH_i(X'')$ (cf. [Kl] §4), by construction compatible with ϵ via the cycle map. In particular, E operates on $A_i(X'')$, and we have

$\mathrm{Ker}\,E = \mathrm{Ker}\,\epsilon \cap A_i(X'') = \mathrm{Ker}(A_i(X'') \xrightarrow{\pi_*\alpha^*} A_i(X))$. Since

$A_i(X'') \xrightarrow{\alpha^*} A_i(X')$ is obviously surjective, and

$A_i(X') \xrightarrow{\pi^*} A_i(X')$ is surjective by similar reasoning as in the proof of 7.6, we get $\operatorname{Im} E \cong \operatorname{Im} \pi_*\alpha^* \cong A_i(X)$. If $\varphi_X \otimes \mathbb{Q}_\ell$ is an isomorphism, it induces an isomorphism

$(\operatorname{Im} E) \otimes \mathbb{Q}_\ell \cong \operatorname{Im} \epsilon \cong H_{2i}(X,\mathbb{Q}_\ell(i))^{G_k}$, which shows the claim.

<u>7.16. Remark</u> In short terms, the Tate conjecture for $X'' \times X''$ assures that the decomposition of $H_{2i}(X'')$ into $W_0H_{2i}(X)$ and $\operatorname{Ker} \pi_*\alpha^*$ is algebraic.

<u>7.17. Theorem</u> Let X have a good proper cover X'' for $H_*^{ét}$. If

i) Tate C) is true for X'' and dimension i, and

ii) for no composition factor W of $H_c^{2i}(\overline{X}'',\mathbb{Q}_\ell)$ the L-function of W has a zero at $s = i + \dim_a k$ (e.g., if char $k = p > 0$), and

iii) Tate B) is true for X and X'', for dimension i,

then Tate C) is true for X and dimension i.

<u>Proof</u> First we remark that condition ii) is generally conjectured for the L-functions of motives which are pure of weight $2i$ (cf. [D6] p. 319, conj. (a)), and holds for positive characteristic by Deligne's fundamental result in [D9] and Grothendieck's formula

$$\prod_{y \in |S|} \frac{1}{\det(1-\mathrm{Fr}_y Ny^{-s}|\mathcal{F}_y)} = \prod_{j=0}^{2\dim S} \det(1 - \mathrm{Fr}\, p^{-s}|H_c^j(\overline{S},\mathcal{F}))^{(-1)^{j+1}}$$

for a constructible $\mathbb{Q}_\ell$-sheaf $\mathcal{F}$ on S (see [SGA 4 $\frac{1}{2}$]

[rapport] 3.1). Here $\bar{S} = S \times_{\mathbb{F}_p} \bar{\mathbb{F}}_p$, $\bar{\mathcal{F}}$ is the pull-back of $\mathcal{F}$ to $\bar{S}$, and Fr is the absolute Frobenius over $\mathbb{F}_p$. If $\mathcal{F}$ is mixed of weights $\leq w$, then by [D9] 3.3.4 $H_c^j(\bar{S},\bar{\mathcal{F}})$ is mixed of weights $\leq w + j$, so that $\det(1 - \mathrm{Fr}\, p^{-s} | H_c^j(\bar{S},\bar{\mathcal{F}}))$ has at most zeroes for $\mathrm{Re}(s) = \frac{m}{2}$, $m \in \mathbb{Z}$ with $m \leq w + j$ (for an eigenvalue α of weight m of Fr one has $\alpha p^{-s} = 1$ at most for $\mathrm{Re}(s) = \frac{m}{2}$).

Instead of iii) we shall only assume

iii′) $A_i(Y) \otimes \mathbb{Q}_\ell \longrightarrow H_{2i}(\bar{Y},\mathbb{Q}_\ell(i))^{G_k}$ is injective for $Y = X, X''$.

For a subquotient W of $H_c^{2i}(\bar{X}'',\mathbb{Q}_\ell)(i)$ let $c(W)$ be the pole order of $L(W,s)$ at $s = \dim_a k$. Then assumption ii) implies

$$c(W) \geq \dim_{\mathbb{Q}_\ell} W_{G_k},$$

where W_{G_k} is the module of coinvariants. let

$K_\ell = \mathrm{Ker}\, \pi_* \alpha^* \subseteq H_{2i}(\overline{X''},\mathbb{Q}_\ell(i))$ and

$K = \mathrm{Ker}(A_i(X'') \xrightarrow{\pi_*\alpha^*} A_i(X))$, then by assumption we have a direct sum decomposition of G_k-modules

$$H_{2i}(\overline{X''},\mathbb{Q}_\ell(i)) = W_0 H_{2i}(\bar{X},\mathbb{Q}_\ell(i)) \oplus K_\ell,$$

and, by dualizing, a decomposition

$$H_c^{2i}(\overline{X''},\mathbb{Q}_\ell(i)) = H_c^{2i}(\bar{X},\mathbb{Q}_\ell(i))/W_{-1} \oplus \check{K}_\ell.$$

By the assumption ii) and iii)′ we get the following system of equalities and inequalities:

$$\begin{array}{ccccc}
c(H_c^{2i}(\overline{X''},\mathbb{Q}_\ell(i))) & = & c(H_c^{2i}(\overline{X},\mathbb{Q}_\ell(i)/W_{-1}) & + & c(\check{K}_\ell) \\
\vee\!\!/ & & \vee\!\!/ & & \vee\!\!/ \\
\dim H_c^{2i}(\overline{X},\mathbb{Q}_\ell(i))_{G_k} & = & \dim\,(H_c^{2i}(\overline{X},\mathbb{Q}_\ell(i))/W_{-1})_{G_k} & + & \dim(\check{K}_\ell)_{G_k} \\
\| & & \| & & \| \\
\dim H_{2i}(\overline{X''},\mathbb{Q}_\ell(i)\,^{G_k} & = & \dim W_0H_{2i}(\overline{X},\mathbb{Q}_\ell(i))^{G_k} & + & \dim K_\ell^{G_K} \\
\vee\!\!/ & & \vee\!\!/ & & \vee\!\!/ \\
\mathrm{rk}\,A_i(X'') & = & \mathrm{rk}\,A_i(X) & + & \mathrm{rk}\,K
\end{array}$$

If Tate C) is true for X'' and dimension i, the left inequalities are both inequalities, hence we also have equalities on the right. Finally, $c(H_c^{2i}(\overline{X},\mathbb{Q}_\ell(i)) = c(H_c^{2i}(\overline{X},\mathbb{Q}_\ell(i))/W_{-1})$, since the L-function of W_{-1} has no zero or pole at $s = \dim_a k$ by the convergent Euler product formula in this range.

7.18 Remark If is perhaps no surprise that homology is the right setting for cycles on singular varieties, since the Fulton Chow groups form a homology theory. One way wonder whether other theories could work for cohomology, e.g., the groups $H^j_{Zar}(X,K_j)$. However, in a note to the author (2/11/87), Bloch has given an argument that there is no contravariant Chow theory coinciding with $CH^*(X)$ for smooth X and generating all Hodge or Tate cycles in the cohomology of singular varieties (see appendix A).

§8. Homology and K-theory of singular varieties

8.1. We now proceed to higher K-groups. We fix a field k, a category $\mathcal{V}$ of varieties over k, a field F of characteristic zero, an F-linear tensor category $\mathcal{T}$, and a Poincaré duality theory $(H^*(\ ,*),H_*(\ ,*))$ with weights and values in $\mathcal{T}$.

In the previous section we studied the cycle map

$$cl_i : CH_i(X) \longrightarrow \Gamma H_{2i}(X,i)$$

on the Fulton Chow group $CH_i(X)$. If X is smooth of pure dimension d, then there is an isomorphism

$$Ch_i(X)\otimes\mathbb{Q} \cong CH^{d-i}(X)\otimes\mathbb{Q} \cong K_0(X)^{(d-i)} = H^{2(d-i)}_{\mathcal{M}}(X,\mathbb{Q}(d-i))$$

by the Grothendieck-Riemann-Roch theorem [SGA 6] XIV 4, and the cycle map can be defined via Chern characters on K_0. In general, we have an isomorphism

$$CH_i(X) \otimes \mathbb{Q} \cong K_0'(X)^{(-i)} = H^{\mathcal{M}}_{2i}(X,\mathbb{Q}(i)) ,$$

where $K_0'(X)$ is the Grothendieck group of coherent sheaves on X, cf. 6.12.4 a). If there is a reasonable theory of Chern classes for our cohomology theory, cl_i again can be expressed by a kind of "Chern character" on K_0', and this can be extended to Quillen's higher K'-groups. Namely, Gillet [Gi] has written down a set of axioms for a Poincaré duality theory assuring the following (cf. also Schechtman's paper [Sche], and [Bei 1] 2.3):

a) There is a relative Chern character on higher K-groups

$$ch^Z : K_m^Z(X) \longrightarrow \bigoplus_{j\geq 0} \Gamma H_Z^{2j-m}(X,j),$$

compatible with the contravariance for morphisms $(X',Z') \longrightarrow (X,Z)$ as in 6.1 a).

b) If X is embeddable in a smooth variety M of pure dimension N (e.g., if X is quasi-projective), the map τ defined by the commutative diagram

$$\begin{array}{ccc} K'_m(X) & \xrightarrow{\tau} & \bigoplus_{b\in\mathbb{Z}} \Gamma H_{m-2b}(X,b) \\ \int \uparrow & & \uparrow Td(M)\cap \\ K^X_m(M) & \xrightarrow{ch^X} & \bigoplus_{j\in\mathbb{Z}} \Gamma H^{2j-m}_X(M,j) \end{array} \tag{8.1.1}$$

is independent of M and compatible with the covariance (resp. contravariance) for proper morphisms (resp. étale morphisms) of K'_m and homology. Here

$Td(M) \in \bigoplus_{\upsilon\in\mathbb{Z}} \Gamma H_{2\upsilon}(M,\upsilon) \cong \bigoplus_{\upsilon\in\mathbb{Z}} \Gamma H^{2N-2\upsilon}(M,N-\upsilon)$

is the Todd class of M [SGA 6], and the isomorphism $K'_m(X) \cong K^X_m(M)$ for smooth M is the one proved by Quillen [Q1], it coincides with the cappropduct with $[\mathcal{O}_M] \in K'_0(M)$.

c) ch and τ satisfy the compatibilities as in the Riemann-Roch theorem of Baum, Fulton and Mac Pherson, generalized to higher algebraic K-theory (see [Gi] 4 .1).

In fact, the morphisms of functors

$$(ch,\tau) : (K_*(\ ),K'_*(\ )) \longrightarrow (\Gamma H^*(\ ,*),\Gamma H_*(\ ,*))$$

is compatible with all the functorialities of twisted Poincaré duality theories, except that the fundamental class of K-theory, $[\mathcal{O}_X]$, is <u>not</u> mapped to the fundamental class η_X. This is closely related to the Riemann-Roch theorem with denominators; for smooth X one has $\tau([\mathcal{O}_X]) = Td(X)$.

A compatibility, which is not explicitly stated in [Gi], but which will be of some importance for us, is the commuting of the Chern character with the relative sequence for K-theory and for cohomology (6.1 c)). Namely, for $Z \subset Y \subset X$ closed immersions one has a commutative diagram

$$
\begin{array}{lccccccc}
 & \ldots \to K^{Z}_{2j-1}(X) & \to & K^{Y}_{2j-i}(X) & \longrightarrow & K^{Y-Z}_{2j-i}(X-Z) & \xrightarrow{\delta} & K^{Z}_{2j-i-1}(X) \to \ldots \\
(8.1.2) & \big\downarrow ch^{Z} & (1) & \big\downarrow ch^{Y} & (2) & \big\downarrow ch^{Y-Z} & (3) & \big\downarrow \\
 & \ldots \to \Gamma H^{i}_{Z}(X,j) & \to & \Gamma H^{i}_{Y}(X,j) & \longrightarrow & \Gamma H^{j}_{Y-Z}(X-Z,j) & \xrightarrow{\delta} & \Gamma H^{i+1}_{Z}(X,j) \to \ldots
\end{array}
$$

The commuting of (2) is the functoriality 8.1 a) above, and the commuting of (1) and (3) follows from the fact that ch^{Z} is induced by a map between the two homotopy fibrations, whose long exact homotopy sequences give rise to the top and bottom sequences in 8.1.2, cf. [Gi] 2.34 ii).

Similarly, for $Y \subset X$ and $U = X - Y \subset X$ the open complement, we have a commutative diagram

$$
\begin{array}{lccccccc}
 & \ldots \to K'_{m}(Y) & \longrightarrow & K'_{m}(X) & \longrightarrow & K'_{m}(U) & \xrightarrow{\delta} & K'_{m-1}(Y) \to \ldots \\
(8.1.3) & \tau_{Y} \big\downarrow & (1)' & \tau_{X} \big\downarrow & (2)' & \tau_{U} \big\downarrow & (3)' & \tau_{Z} \big\downarrow \\
 & \ldots \to \Gamma H_{m-2b}(Y,b) & \to & \Gamma H_{m-2b}(X,b) & \to & \Gamma H_{m-2b}(U,b) & \xrightarrow{\delta} & \Gamma H_{m-1-2b}(Y,b) \to \ldots
\end{array}
$$

The commuting of (1)' and (2)' follows from 8.1b), and the commutativity of (3)' is the commutativity of

$$
\begin{array}{rccc}
 & K'_{m}(U) & \longrightarrow & K'_{m-1}(Y) \\
 & \int\big\uparrow & & \int\big\uparrow \\
 & K^{U}_{m}(M^{0}) & \longrightarrow & K^{Y}_{m-1}(M) \\
ch^{U} & \big\downarrow & & \big\downarrow ch^{Y} \\
 & \bigoplus_{j\in\mathbb{Z}} \Gamma H^{2j-m}_{U}(M^{0},j) & \xrightarrow{\delta} & \bigoplus_{j\in\mathbb{Z}} \Gamma H^{2j-m+1}_{Y}(M,j) \\
\cap\, Td(M^{0}) & \big\downarrow & & \big\downarrow \cap\, Td(M) \\
 & \bigoplus_{b\in\mathbb{Z}} \Gamma H_{m-2b}(U,b) & \xrightarrow{\delta} & \bigoplus_{b\in\mathbb{Z}} \Gamma H_{m-1-2b}(Y,b),
\end{array}
$$

which follows from the compatibility of the capproduct with δ (this is implied by [Gi] 1.2 vii)) and the fact that $Td(M)$ restricts to $Td(M^0)$. Here we have embedded X in a smooth variety M, and let $M^0 = M \cap U = M \setminus Y$.

<u>8.2. Remark</u> Our general setting forces us to apply Γ for obtaining abelian groups from the objects of $\mathcal{T}$, and Γ is usually not exact so that $(\Gamma H^*(,*),\Gamma H_*(,*))$ will not form a twisted Poincaré duality theory. However, in the cases we are interested in there is an exact, faithful tensor-functor

$$v : \mathcal{T} \longrightarrow \underline{\mathrm{Vec}}_F$$

into the category of (finite-dimensional) vector spaces over F (compare remark 6.2a)) with $v(1) = F$. Then $(vH^*(,*),vH_*(,*))$ is a twisted Poincaré duality theory, and, since $\Gamma(A) = \mathrm{Hom}_{\mathcal{T}}(1,A) \subseteq \mathrm{Hom}_F(F,vA) \cong vA$ is an injection, we may regard all elements in $\Gamma(A)$ as elements in vA (the "underlying F-vector space") which happen to lie in the subspace ΓA. In particular, we may regard (ch,τ) as morphisms into $(vH^*(,*),vH_*(,*))$, and they are morphisms of twisted Poincaré duality theories, except that the compatibility between the fundamental classes $[\mathcal{O}_X]$ and η_X is missing.

<u>8.3. Examples</u>, where these "Riemann-Roch transformations" exist, are

a) the ℓ-adic theory (6.7 and 6.8, [Gi] 1.4 iii)),
b) the Hodge theory (6.9, [Gi] 1.4 iv)),
c) The deRham theory (6.10, [Gi] 1.4 i), no weights),
d) realizations for absolute Hodge cycles (6.11, by combining

a) - c)).

The fact that the images under ch and τ lie in the Γ-subspaces and are compatible under the comparison isomorphisms can be checked for the universal Chern classes $c_{i,n} \in H^{2i}(B.GL_n/k,i)$, where it is clear from the construction and the known case of the first Chern class. One may also use absolute theories (étale cohomology of X over k (instead of $\bar{k}$), Deligne cohomology,...) and the restriction maps into the geometric theories.

8.4. In the above examples, ch and τ give rise to actual morphisms of twisted Poincaré dualities

(8.4.1)
$$r : H^i_{\mathcal{M},Z}(X,\mathbb{Q}(j)) \longrightarrow vH^i_Z(X,j),$$
$$r' : H^{\mathcal{M}}_a(X,\mathbb{Q}(b)) \longrightarrow vH_a(X,b),$$

since, by the properties of Chern characters, ch^Z restricted to $H^i_{\mathcal{M},Z}(X,\mathbb{Q}(j)) = K^Z_{2j-i}(X)^{(j)}$ has the only non-trivial component in $vH^i_Z(X,j)$ and similarly $\tau(H^{\mathcal{M}}_a(X,\mathbb{Q}(b))) \subseteq vH_a(X,b)$, cf. [J2] 3.6b). Moreover, since for smooth X we have
$Td(X) = \tau([\mathcal{O}_X]) = \eta_X + \text{terms in} \bigoplus_{v>\dim X} vH_{2v}(X,v)$,
and since there is a commutative diagram

(8.4.2)
$$\begin{array}{ccc} & \overset{\tau}{\nearrow} & \bigoplus_{i\geq 0} CH_i(X)\otimes\mathbb{Q} \\ K'_0(X) & & \downarrow cl \\ & \underset{\tau}{\searrow} & \bigoplus_{i\geq 0} vH_{2i}(X,i), \end{array}$$

we have $r'(\eta^{\mathcal{M}}_X) = \eta_X$ for the motivic fundamental class $\eta^{\mathcal{M}}_X = [X] \in CH_{\dim X}(X)\otimes \mathbb{Q} \cong K'_0(X)^{(-\dim X)} \cong H^{\mathcal{M}}_{2\dim X}(X,\mathbb{Q}(\dim X))$. Here the upper τ in 8.4.2 is Riemann-Roch transformation for

Chow theory ([Fu] and [Gi] 4.1 and 1.4 ii)), which is known to induce inverses for the isomorphisms in 6.12.4a).

In generalization of the conjectures in §5 and §7 we propose to study the images of the maps

$$(8.4.3)\qquad r'_{a,b}\otimes F : H^{\mathcal{M}}_a(X,\mathbb{Q}(b)) \otimes_{\mathbb{Q}} F \longrightarrow \Gamma H_a(X,b).$$

<u>8.5. Conjecture</u> If k is a global or finite field and X is an algebraic k-scheme, then for $a,b \in \mathbb{Z}$ the map

$$r'_{a,b} \otimes \mathbb{Q}_\ell : H^{\mathcal{M}}_a(X,\mathbb{Q}(b)) \otimes \mathbb{Q}_\ell \longrightarrow H^{\text{ét}}_a(\overline{X},\mathbb{Q}_\ell(b))^{G_k}$$

is surjective, provided $\ell \neq \operatorname{char} k$.

<u>8.6. Conjecture</u> If X is a variety over $\mathbb{C}$ that is defined over a number field, then

$$r'_{a,b} : H^{\mathcal{M}}_a(X,\mathbb{Q}(b)) \longrightarrow \Gamma_{*}(H_a(X(\mathbb{C}),\mathbb{Q}(b)))$$

is surjective for all $a,b \in \mathbb{Z}$.

<u>8.7. Conjecture</u> If X is a variety over a number field k, then

$$r'_{a,b} : H^{\mathcal{M}}_a(X,\mathbb{Q}(b)) \longrightarrow \Gamma_{\mathcal{A}*}(H^{AH}_a(X,b))$$

is surjective for all $a,b \in \mathbb{Z}$.

<u>8.8 Remark</u> As explained above, these conjectures are equivalent to the surjectivity of

$$\tau_{a,b} \otimes \mathbb{Q}_\ell \; : \; K'_{b-2a}(X) \otimes \mathbb{Q}_\ell \longrightarrow H_a^{ét}(\overline{X},\mathbb{Q}_\ell(b))^{G_k}$$

$$\tau_{a,b} \otimes \mathbb{Q} \; : \; K'_{b-2a}(X) \otimes \mathbb{Q} \longrightarrow (2\pi\sqrt{-1})^{-b} W_{-2b} H_a(X(\mathbb{C}),\mathbb{Q}) \cap F^{-b}$$

$$\tau_{a,b} \otimes \mathbb{Q} \; : \; K'_{b-2a}(X) \otimes \mathbb{Q} \longrightarrow \Gamma_{\mathcal{A}\mathcal{H}}(H_a^{AH}(X,b)).$$

The next result shows, that at least for projective X these conjectures are implied by the conjectures 5.18, 5.19 and 5.20 (for smooth schemes). Since we are interested in the general principle, we just assume that we have morphisms of twisted Poincaré duality theories as in 8.4.1, and consider the surjectivity of $r'_{a,b}$ as in 8.4.3, which for smooth X of pure dimension d by Poincaré duality agrees with the map

$$(r = ch) \otimes F \; : \; H_{\mathcal{M}}^{2d-a}(X,\mathbb{Q}(d-b))\otimes F \longrightarrow \Gamma H^{2d-a}(X,d-b)$$

studied in §5. Consider the following properties for our Poincaré duality theory.

n) (projective bundle isomorphism) The capproduct induces an isomorphism

$$\bigoplus_{\upsilon=0}^{n} H^{2n-a-2\upsilon}(\mathrm{Spec}\ k, n-b-\upsilon) \xrightarrow[\sim]{\oplus_\upsilon p^* \cap \xi^\upsilon} H_a(\mathbb{P}_k^n, b)$$

for $a,b \in \mathbb{Z}$, where $\xi^\upsilon \in \Gamma H_{2n-2\upsilon}(\mathbb{P}_k^n, n-\upsilon)$ is the cycle class of H^υ for a hyperplane H of $\mathbb{P}_k^n$ and $p : \mathbb{P}_k^n \longrightarrow \mathrm{Spec}\ k$ is the projection.

o) (geometric chomology) One has $H^i(\mathrm{Spec}\ k,j) = 0$ for $i \neq 0$, and $\Gamma H^0(\mathrm{Spec}\ k,0) = F$.

Usually n) is one of the conditions needed to obtain a theory of Chern classes as above, and both n) and o) hold for the examples in 8.3.

8.9. Lemma Assume that n) and o) are satisfied. Let $X \xrightarrow{i} \mathbb{P}^n_k$ be a closed subvariety with open complement $U = \mathbb{P}^n_k \setminus X$, and let a,b be integers. If $r'_{a+1,b} \otimes F$ is surjective for U, then $r'_{a,b} \otimes F$ is surjective for X.

Proof Step 1 From n), o) and 6.6 we get:

$$H_a(\mathbb{P}^n_k,b) = 0 \quad \text{for odd } a,$$
$$\Gamma H_a(\mathbb{P}^n_k,b) = 0 \quad \text{for } a \neq 2b,$$

$r_{2b,b} \otimes F$ is an isomorphism for $\mathbb{P}^n_k$ and all b. Here we used the fact that $H^{\#}_{2b}(\mathbb{P}^n_k,\mathbb{Q}(b)) \cong CH_b(\mathbb{P}^n_k) \otimes \mathbb{Q} \cong \mathbb{Q}$ for $b = 0,\ldots,n$ with the class of H^{n-b} as basis.

Step 2 For even $a = 2c$, $0 \leq a \leq 2d$, $d = \dim X$, the morphisms

$$H_a(X,b) \xrightarrow{i_*} H_a(\mathbb{P}^n_k,b)$$

are surjective. In fact, by n) every element of $H_a(\mathbb{P}^n_k,b)$ is of the form $\alpha \cap \xi^{n-c}$, and by the projection formula it suffices to prove that ξ^{n-c} is in the image of i_*. But one has a commutative diagram

$$\begin{array}{ccc} \Gamma H_{2c}(X,c) & \xrightarrow{i_*} & \Gamma H_{2c}(\mathbb{P}^n k,c) \\ r'\uparrow & & \uparrow r' \\ H^{\mathcal{M}}_{2c}(X,\mathbb{Q}(c)) & \longrightarrow & H^{\mathcal{M}}_{2c}(\mathbb{P}^n_k,\mathbb{Q}(c)), \end{array}$$

so we only have to prove that the class of H^{n-c} is in the image of the map below, which is well known (for $c = d$ one has $i_*\eta_X = \deg X \cdot [H^{n-c}] \neq 0$, and for $0 \leq c \leq d$ one may use the projection formula again).

<u>Step 3</u> Let a be even, $0 \leq a \leq 2d$, then by step 1 we have an exact sequence

$$0 \to H_{a+1}(U,b) \to H_a(X,b) \longrightarrow H_a(\mathbb{P}^n_k,b)$$

and hence a commutative exact diagram

$$\begin{array}{ccccccc} 0 \longrightarrow & \Gamma H_{a+1}(U,b) & \longrightarrow & \Gamma H_0(X,b) & \longrightarrow & \Gamma H_a(\mathbb{P}^n_k,b) \\ & \uparrow F\otimes r'_U & & \uparrow F\otimes r'_X & & \uparrow F\otimes r'_{\mathbb{P}^n} \\ & F \otimes H^{\mathcal{M}}_{a+1}(U,\mathbb{Q}(b)) & \to & F \otimes H^{\mathcal{M}}_a(X,\mathbb{Q}(b)) & \xrightarrow{i_*} & F \otimes H^{\mathcal{M}}_a(\mathbb{P}^n_k,\mathbb{Q}(b)), \end{array}$$

where either $\Gamma H_a(\mathbb{P}^n_k,b) = 0$, or else $a = 2b$ and $F \otimes r'_{\mathbb{P}^n}$ is an isomorphism and i_* is surjective. This implies the claim.

<u>Step 4</u> If a is odd, $0 < a < 2d$, we get an isomorphism $H_{a+1}(U,b) \xrightarrow{\sim} H_a(Z,b)$, so the claim is clear by functoriality of r'. q.e.d

We can extend the theorems 5.13, 5.15 and 5.17 to arbitrary varieties X. Let X be of (not necessarily pure) dimension d, and let

$$E^2_{d,-d+1}(X) = \mathrm{Ker}\Big(\bigoplus_{x \in X_{(d)}} \kappa(x)^{\times} \xrightarrow{\mathrm{div}} \bigoplus_{x \in X_{(d-1)}} \mathbb{Z} \Big)$$

be the E^2-term of the Quillen spectral sequence 6.12.5 for X. If X is smooth of pure dimension d, we have $E^2_{d,-d+1}(X) = \mathcal{O}(X)^{\times}$, but in general there is only a homomorphism

$$\mathcal{O}(X)^{\times} \longrightarrow E^2_{d,-d+1}(X)$$

which might be neither injective nor surjective.

__8.10 Theorem__ Let X be a variety of dimension d over $\mathbb{C}$, then there is a canonical isomorphisim

$$\varphi : E^2_{d,-d+1}(X) \xrightarrow{\sim} H^{\mathcal{D}}_{2d-1}(X,\mathbb{Z}(d-1)),$$

where $H^{\mathcal{D}}$ is the Deligne homology [Bei 1], [J2]. In particular, there is a surjection

$$\tau : K'_1(X) \longrightarrow \Gamma_{\#}(H_{2d-1}(X,\mathbb{Z}(d-1)))$$

$$= (2\pi\sqrt{-1})^{-d+1}H_{2d-1}(X,\mathbb{Z}) \cap F^{-d+1}H_{2d-1}(X,\mathbb{C})$$

inducing the map $\tau_{2d-1,d}$ in 8.1.1 for Betti homology, and conjecture 8.6 is true for (a,b) = (0,0),(2d-2,d-1), (2d-1,d-1) and (2d,d), hence in general for curves.

__Proof__ If X is smooth of pure dimensin d, φ is the isomorphism

$$\alpha = c_{1,1} : \mathcal{O}(X)^{\times} \longrightarrow H^1_{\mathcal{D}}(X,\mathbb{Z}(1)) \cong H^{\mathcal{D}}_{2d-1}(X,\mathbb{Z}(d-1))$$

used in 5.13. For general (reduced) X, there is a smooth open subvariety $U \subseteq X$ of pure dimension d, such that the complement $Y = X \setminus U$ has dimension strictly less than d. We obtain a commutative diagram

$$(8.10.1)\qquad \begin{array}{ccccc} 0 \to H^{\mathcal{D}}_{2d-1}(X,\mathbb{Z}(d-1)) & \to & H^{\mathcal{D}}_{2d-1}(U,\mathbb{Z}(1)) & \to & H^{\mathcal{D}}_{2d-2}(Y,\mathbb{Z}(d-1)) \\ \uparrow & & \int \uparrow \alpha & & \int \uparrow cl \\ 0 \to E^2_{d,d+1}(X) & \longrightarrow & E^2_{d,-d+1}(U) & \xrightarrow{\text{div}} & \bigoplus_{\substack{x \in X_{(d-1)} \\ x \in Y}} \mathbb{Z} \end{array}$$

inducing the dotted isomorphism we needed, see [J2] 3.1. The second claim follows from the diagrams

$$(8.10.2)\qquad \begin{array}{ccccccc} 0 \to & H_{2d}(X,\mathbb{C})/H_{2d}(X,\mathbb{Z}(d-1))+F^{d+1} & \to & H^{\mathcal{D}}_{2d-1}(X,\mathbb{Z}(d-1)) & \longrightarrow & \Gamma_{*}H_{2d-1}(X,\mathbb{Z}(d-1)) & \to 0 \\ & \uparrow \int & & \int \uparrow \varphi & & \int \uparrow & \\ 0 \longrightarrow & \bigoplus_{x \in X_{(d)}} \mathbb{C}^{\times} & \longrightarrow & E^2_{d,-d+1}(X) & \longrightarrow & E^2_{d,-d+1}(X) / \bigoplus_{x \in X_{(d)}} \mathbb{C}^{\times} & \to 0 \end{array}$$

$$(8.10.3)\qquad \begin{array}{ccc} K_1'(X) & \xrightarrow{\ \tau\ } & \\ \downarrow & & H^{\mathcal{D}}_{2d-1}(X,\mathbb{Z}(d-1)) \\ E^2_{d,-d+1}(X) & \xrightarrow[\sim]{\ \varphi\ } & \end{array}$$

The upper sequence in 8.10.2 is the canonical exact sequence for Deligne homology (see, e.g., [J 2] 1.18 b)), and the factorization and bijectivity of the left vertical map can easily be seen by restricting to U, whereby source and target do not change, and using 5.13. The vertical map in 8.10.3 is induced by the Quillen spectral sequence, since $E^1_{p,q}(X) = 0$

for $p > d = \dim X$. It is surjective by lemma 8.11 below, and if we define τ by commutativity of 8.10.3, the compatibility with $\tau_{2d-1,d-1}$ is proved in [J 2] 3.4.

This shows the surjectivity of $r'_{2d-1,d-1}$ (cf. remark 8.8), and the remaining claims are obvious: the cases $(a,b) = (0,0)$, $(2d-2,d-1)$ and $(2d,d)$ amount to the Hodge conjecture for the known cases of dimension d (trivial), $d-1$ (Lefschetz' theorem), and 0 (trivial), cf. theorem 7.9. Finally, for curves we have $\Gamma H_a(X,b) \neq 0$ at most for $(a,b) = (0,0)$, $(1,0)$ and $(2,1)$ by lemma 8.12 below.

<u>8.11. Lemma</u> (compare [Sou 3] théoréme 4 iv)) For a variety X of dimension d over a field k one has $E^2_{d,-d+1}(X) = E^{\infty}_{d,-d+1}(X)$ in the Quillen spectral sequence 6.12.5, so that the edge morphism $K'_1(X) \to E^2_{d,-d+1}(X)$ is surjective.

<u>Proof</u> We know that all differentials $d^r_{p,q}$ from the line $p + q = 1$ to the line $p + q = 0$ vanish for $r \geq 2$ after tensoring with $\mathbb{Q}$ (Soulé's result used for the proof of 6.12.4 a)). On the other hand we have

$$\mathrm{Tor}\,(E^2_{d,-d+1}(X)) = \bigoplus_{x \in X_{(d)}} \mu_{\kappa(x)} = \bigoplus_{x \in X_{(d)}} \mu_{k_x},$$

where μ_L is the group of roots of unity in a field L and k_x is the algebraic closure of k in $\kappa(x)$. Let $\tilde{X}$ be the normalization of X, then the commutative diagram

$$\begin{array}{ccccc}
\bigoplus_{X_{(d)}} K_1(k_x) & \longrightarrow & K_1'(\tilde{X}) & \longrightarrow & K_1'(X) \\
\| & & \downarrow & & \downarrow \\
\bigoplus_{x\in X_{(d)}} E^2_{0,1}(\mathrm{Spec}\ k_x) & \longrightarrow & E^2_{d,-d+1}(\tilde{X}) & \longrightarrow & E^2_{d,-d+1}(X) ,
\end{array}$$

obtained from the covariance of Quillen's spectral sequence for finite morphisms and its contravariance for flat morphisms, shows that $\bigoplus_{x\in X_{(d)}} \mu_{k_x}$ lies in the image of the right vertical map, hence the differentials also vanish on the torsion subgroup of $E^r_{d,-d+1}(X)$ for $r \geq 2$.

<u>8.12. Lemma</u> For a twisted Poincaré duality theory with weights one has $\Gamma H_a(X,b) \neq 0$ at most for $a \geq 2b \geq 0$ and $b \geq a - d$, $d = \dim X$, hence in the triangle indicated below:

(8.12.1) 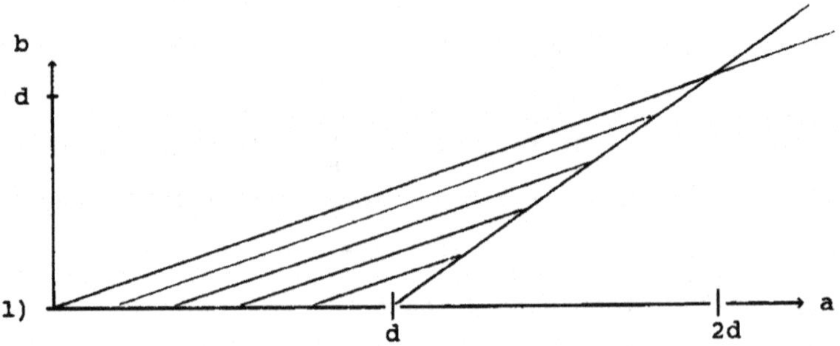

<u>Proof</u> (compare 5.11) In view of 6.5 we must have

$$2b - a \leq 0 \leq 2b \quad \text{and} \quad a \leq d, \text{ or}$$

$$2b - a \leq 0 \leq 2b - 2(a - d) \quad \text{and} \quad a \geq d.$$

Note, that for smooth X of pure dimension d the diagram 8.12.1 is transformed into 5.11.1 via $(a,b) \mapsto (2d - a, d - b)$, by using Poincaré duality.

8.13. Theorem Let X be a variety of dimension d over a finitely generated field k, and let $\ell \neq \mathrm{char}\, k$ be a prime.

a) There are canonical homomorphisms for all $n \geq 1$

$$\varphi : E^2_{d,-d+1}(X)/\ell^n \longrightarrow H^{ét}_{2d-1}(X,\mathbb{Z}/\ell^n(d-1)),$$

compatible with the transition maps for different n, such that the induced map

$$\varphi : E^2_{d,-d+1}(X)\hat{} = \varprojlim_n E^2_{d,-d+1}(X)/\ell^n \to H^{ét}_{2d-1}(X,\mathbb{Z}_\ell(d-1))$$

is an isomorphism.

b) The restriction

$$H^{ét}_{2d-1}(X,\mathbb{Z}_\ell(d-1)) \longrightarrow H^{ét}_{2d-1}(\overline{X},\mathbb{Z}_\ell(d-1))^{G_k}$$

is surjective.

c) The induced map

$$K'_1(X) \longrightarrow E^2_{d,-d+1}(X) \longrightarrow H^{ét}_{2d-1}(\overline{X},\mathbb{Q}_\ell(d-1))^{G_k}$$

coincides with $\tau_{2d-1,d-1}$.

In particular, conjecture 8.5 is true for $(a,b) = (0,0)$, $(2d-1,d-1)$ and $(2d,d)$, hence for curves in general.

Proof This follows as in 8.10: Choosing U as there, the φ_n are defined by the commutative exact diagram

$$
\begin{array}{lllll}
0 \to H^{ét}_{2d-1}(X,\mathbb{Z}/\ell^n(d-1)) & \to & H^{ét}_{2d-1}(U,\mathbb{Z}/\ell^n(d-1)) & \to & H^{ét}_{2d-2}(Y,\mathbb{Z}/\ell^n(d-1)) \\
(8.13.1) \quad \uparrow \varphi_n & & \uparrow 5.15 & & \uparrow cl \\
0 \longrightarrow E^2_{d,-d+1}(X) & \longrightarrow & E^2_{d,-d+1}(U) & \xrightarrow{\ div\ } & \bigoplus_{x \in Y_{(d-1)}} \mathbb{Z},
\end{array}
$$

where we have used that $H^{ét}_a(Y,\mathbb{Z}/\ell^n(b)) = 0$ for $a > 2d - 2 \geq 2 \dim Y$ (cf. 8.13.3 below). Passing to the inverse limit over n in the top row and the pro-ℓ-completions in the bottom row, we obtain a commutative diagram

$$
\begin{array}{lllll}
0 \to H^{ét}_{2d-1}(\overline{X},\mathbb{Z}_\ell(d-1)^{G_k} & \to & H^{ét}_{2d-1}(\overline{U},\mathbb{Z}_\ell(d-1))^{G_k} & \longrightarrow & H^{ét}_{2d-2}(\overline{Y},\mathbb{Z}_\ell(d-1))^{G_k} \\
\qquad \uparrow res_X & & \uparrow res_U & & \wr \uparrow res_Y \\
0 \to H^{ét}_{2d-1}(X,\mathbb{Z}_\ell(d-1) & \to & H^{ét}_{2d-1}(U,\mathbb{Z}_\ell(d-1)) & \longrightarrow & H^{ét}_{2d-2}(Y,\mathbb{Z}_\ell(d-1)) \\
(8.13.2) \uparrow \varphi & & \wr \uparrow 5.15a) & & \wr \uparrow cl \\
0 \to E^2_{d,-d+1}(X)^\wedge & \to & E^2_{d,-d+1}(U)^\wedge & \longrightarrow & \bigoplus_{x \in Y_{(d-1)}} \mathbb{Z}_\ell
\end{array}
$$

with exact tows: the top row is exact by the left-exactness of taking fixed modules, the middle one by the left-exactness of passing to the inverse limit, and the exactness of the bottom row follows from 8.13.1 and the fact that $\bigoplus_{x \in Y_{(d-1)}} \mathbb{Z}$ is finitely generated torsion-free. The bijectivity of res_Y follows from the fact that $H^{ét}_a(\overline{Y},\mathbb{Z}/\ell^n(b)) = 0$ for $a > 2 \dim Y$, via the Hochschild-Serre spectral sequence

$$(8.13.3) \quad E_2^{p,q} = H^p(G_k, H^{ét}_{-q}(\overline{Y},\mathbb{Z}/\ell^n(b))) \Rightarrow H^{ét}_{-p-q}(Y,\mathbb{Z}/\ell^n(b)).$$

The isomorphism

$$(8.13.4) \qquad \bigoplus_{y \in Y_{(a)}} \mathbb{Z}/\ell^n \xrightarrow[\sim]{cl} H^{ét}_{2a}(Y,\mathbb{Z}/\ell^n(a)), \quad a \geq \dim Y,$$

is well-known; via the relative homology sequence (cf. 6.1 f)) and Poincaré duality it can easily be reduced to the statement that canoincally $\mathbb{Z}/\ell^n \xrightarrow{\sim} H^0_{ét}(Y,\mathbb{Z}/\ell^n)$ for Y smooth and connected. Putting this together we obtain the statements a) and b), since res_U is surjective, as we have seen in the proof of 5.15 (together with 5.16 b)).

The first claim in c) follows as in 8.10: by compatibility of τ and the other maps involved with restriction to U and with Poincaré duality, it suffices to remark that

$$K_1(U) \xrightarrow{\det} \mathcal{O}(U)^\times \xrightarrow{5.15a)} H^1(U,\mathbb{Z}/\ell^n(1) \longrightarrow H^1(\bar{U},\mathbb{Z}/\ell^n(1))^{G_k}$$

is the first Chern class $c_{1,1}$. For the following, diagram 8.10.2 is replaced by the diagram

$$\begin{array}{ccccccccc} 0 \to & H^1_{cont}(G_k,H_{2d}(\bar{X},\mathbb{Z}_\ell(d-1)) & \to & H_{2d-1}(X,\mathbb{Z}_\ell(d-1)) & \to & H_{2d-1}(\bar{X},\mathbb{Z}_\ell(d-1))^{G_k} & \to 0 \\ (8.13.5) & \uparrow\wr & & \uparrow\varphi & & \wr\uparrow & \\ 0 \longrightarrow & \bigoplus_{x\in X_{(d)}} (k_x^\times)^\wedge & \longrightarrow & E^2_{d,-d+1}(X)^\wedge & \longrightarrow & (E^2_{d,-d+1}(X)/\bigoplus_{x\in X_{(d)}} k_x^\times)\otimes\mathbb{Z}_\ell & \to 0, \end{array}$$

in which the left vertical map comes form the isomorphisms

$$H^1_{cont}(G_k,H_{2d}(\bar{X},\mathbb{Z}_\ell(d-1))) \cong H^1_{cont}(G_k,\bigoplus_{x\in\bar{X}_{(d)}} \mathbb{Z}_\ell(1))$$

$$= \bigoplus_{x\in X_{(d)}} H^1_{cont}(G_{k_x},\mathbb{Z}_\ell(1)) \xleftarrow{\sim} \bigoplus_{x\in X_{(d)}} (k_x^\times)^\wedge$$

induced by Kummer theory (cf. 5.16 b); notations as in the proof of 8.11). The exactness of the lower sequence in 8.13.5 follows from the fact hat $E^2_{d,-d+1}(X)/\bigoplus_{x\in X_{(d)}} k_x^\times$ is a finitely generated, torsion-free group, which via the exact sequence

$$0 \longrightarrow E^2_{d,-d+1}(X)/\bigoplus_{x\in X_{(d)}} k_x^\times \longrightarrow E^2_{d,-d+1}(U)/\bigoplus_{x\in U_{(d)}} k_x^\times \longrightarrow \bigoplus_{x\in Y_{(d-1)}} \mathbb{Z}$$

follows from the corresponding statement for U, proved in the proof of 5.15.

Hence the composition

$$K_1'(X) \otimes \mathbb{Z}_\ell \twoheadrightarrow E^2_{d,-d+1}(X) \otimes \mathbb{Z}_\ell \twoheadrightarrow H_{2d-1}(\bar{X},\mathbb{Z}_\ell(d-1))^{G_k}$$

is surjective, and by tensoring with $\mathbb{Q}_\ell$ we obtain the claim of conjecture 8.5 for $(a,b) = (2d-1,d-1)$. The cases $(a,b) = (0,0)$ or $(2d,d)$ again are trivial cases, this time of the Tate conjecture. Since we do not want to assume resolutions of singularities, we cannot apply 7.10 but proceed directly instead. The case $(2d,d)$ follows from the isomorphism

$$\bigoplus_{x\in X_{(d)}} \mathbb{Z}_\ell \xrightarrow[\sim]{cl} H^{et}_{2d}(\bar{X},\mathbb{Z}_\ell(d))^{G_k}$$

used above. For $(0,0)$ we proceed by induction on $d = \dim X$. We know the claim for $d = 0$ (above) and $d = 1$ (by 7.10; we have resolution of singularities and 7.8 b) for $i = 0$). For $d > 1$, choose $U \subsetneq X$ as above, and affine. Then we have an exact sequence

$$\begin{array}{ccccccc} H_1(\bar{U},\mathbb{Z}_\ell(0)) & \to & H_0(\bar{Y},\mathbb{Z}_\ell(0)) & \to & H_0(\bar{X},\mathbb{Z}_\ell(0)) & \to & H_0(\bar{U},\mathbb{Z}_\ell(0)) \\ \wr\| & & & & & & \wr\| \\ H^{2d-1}(\bar{U},\mathbb{Z}_\ell(d)) = 0 & & & & & & 0 = H^{2d}(\bar{U},\mathbb{Z}_\ell(d)) \end{array}$$

since $H^i(\bar{U},\mathbb{Z}_\ell(j)) = 0$ for U affine and $i > \dim U$. This reduces the question to Y and hence to smaller dimension.

§9. Extension classes, algebraic cycles, and the case $i = 2j - 1$

The interesting feature of mixed structures is the existence of non-trivial extensions - in particular those belonging to the weight filtration.

9.0. For example, let as before $(H^*(\ ,*)\ ,H_*(\ ,*))$ be a $\mathcal{T}$-valued twisted Poincaré duality theory with weights, let X be a variety, and let z be a cycle of dimension i on X, supported on a closed subvariety Z of the same dimension. If we assume

p) (cf. [BO] 7.1.1) $H_a(Z,b) = 0$ for $a > 2 \dim Z$

(by 6.5 a) this is always true after tensoring with the field F in the definition 6.5, and it is true for the examples in § 6), then we get an exact sequence

$$(9.0.1)\quad 0 \longrightarrow H_{2i+1}(X,i) \longrightarrow H_{2i+1}(U,i) \longrightarrow H_{2i}(Z,i) \xrightarrow{\delta} H_{2i}(X,i)\ ,$$

with $U = X \setminus Z$. The cycle class $cl(z)$ on Z is an element in $\Gamma H_{2i}(Z,i) = \mathrm{Hom}_{\mathcal{T}}(1,H_{2i}(Z,i))$, and if z is homologically equivalent to zero on X, then by definition $\delta(c\ell(z)) = 0$. In this case z via pull-back gives rise to an extension

$$(9.0.2)\quad 0 \longrightarrow H_{2i+1}(X,i) \longrightarrow E \longrightarrow 1 \longrightarrow 0\ ,$$

and thus defines an element $c\ell'(z)$ in

$$\mathrm{Ext}^1_{\mathcal{T}}(1, H_{2i+1}(X,i)) =: R^1\Gamma H_{2i+1}(X,i) .$$

Let $Z_i(X)_0$ be the subgroup of $Z_i(X)$ formed by the cycles which are homologically equivalent to zero, then the above prescription defines an Abel-Jacobi map

$$(9.0.3) \quad c\ell' : Z_i(X)_0 \longrightarrow R^1\Gamma H_{2i+1}(X,i) ,$$

while the cycle map induces a homomorphism

$$(9.0.4) \quad c\ell : Z_i(X)/Z_i(X)_0 \longrightarrow \Gamma H_{2i}(X,i) .$$

9.1. Remarks a) If X is smooth of pure dimension d, then by Poincaré duality the above corresponds to the diagram

$$(9.1.1) \quad \begin{array}{ccccccccc} 0 & \longrightarrow & H^{2j-1}(X,j) & \longrightarrow & H^{2j-1}(U,j) & \longrightarrow & H^{2j}_Z(X,j) & \longrightarrow & H^{2j}(X,j) \\ & & \Vert & & \cup & & \uparrow z & \nearrow 0 & \\ 0 & \longrightarrow & H^{2j-1}(X,j) & \longrightarrow & E & \longrightarrow & 1 & \longrightarrow & 0 \end{array} ,$$

where $j = d - i$ is now the codimension of the cycle and the fundamental class of z is an element of $\Gamma H^{2j}_Z(X,j)$, and the Abel-Jacobi map can be written as

$$(9.1.2) \quad c\ell' : Z^j(X)_0 \longrightarrow R^1\Gamma H^{2j-1}(X,j) .$$

b) In view of the formula $H^\nu(T,F) = \mathrm{Ext}^\nu_T(\mathbb{Z},F)$ for a sheaf F on a space (site ...) T, it is sometimes suggestive to write $H^\nu(\mathcal{T},A) := \mathrm{Ext}^\nu_{\mathcal{T}}(1,A)$ for an object A of $\mathcal{T}$, while $R^\nu\Gamma$ reminds of taking the ν-th right derivative of $\Gamma(\ ) = \mathrm{Hom}_T(1,\ ) = H^0(T,\ )$. If T has enough injectives, one can in fact compute $\mathrm{Ext}^\nu_T(1,-)$ by injective resolutions; otherwise we have to take the Yoneda Ext-groups.

c) The above relation between cycles and extensions (or torsons) can already be found in [D6] 4.3.

For Hodge theory we recover the classical Abel-Jacobi map.

<u>9.2. Lemma</u> For a mixed Hodge structure H there is a canonical isomorphism

$$(9.2.1)\quad \mathrm{Ext}^1_{MH}(\mathbb{Z},H) \cong W_0H \otimes_{\mathbb{Z}} \mathbb{C}/W_0H + F^0W_0H \otimes_{\mathbb{Z}} \mathbb{C}\ ,$$

and for a smooth proper variety X over $\mathbb{C}$ the homomorphism

$$(9.2.2)\quad Z^j(X)_0 \longrightarrow H^{2j-1}(X,\mathbb{C})/H^{2j-1}(X,\mathbb{Z}(2\pi\sqrt{-1})^j) + F^jH^{2j-1}$$

of 9.1.2 coincides with the classical Abel-Jacobi map. (Here W_0H means the preimage of $W_0H \otimes \mathbb{Q}$ in H , and the quotient in 9.2.1 is to be taken either in the sense that H acts on $H \otimes \mathbb{C}$ or by replacing H by $\mathrm{Im}(H \longrightarrow H \otimes \mathbb{C})$; similarly for 9.2.2).

<u>Proof</u> We show how to deduce this from results of Carlson [Ca], it can also be obtained from Beilinson's papers [Bei 1], [Bei 2]. First note that $H \cong \mathrm{Tor}(H) \oplus H/\mathrm{Tor}(H)$, and $\mathrm{Ext}^1_{MH}(\mathbb{Z},H')$ $= \mathrm{Ext}^1_{Ab}(\mathbb{Z},H') = 0$ for a torsion Hodge structure H' , hence we may assume that H is a lattice. Then, if we write J(H) for the right hand side of 9.2.1, we obtain a morphism of functors

$$(9.2.3)\quad \mathrm{Ext}^1_{MH}(\mathbb{Z},H) \longrightarrow \mathrm{Ext}^1_{MH}(\mathbb{Z},W_0H) \longrightarrow J(W_0H) = J(H)$$

by sending an extension

$$0 \longrightarrow H \xrightarrow{i} E \xrightarrow{\pi} \mathbb{Z} \longrightarrow 0$$

to the extension $0 \to W_0H \to W_0E \to \mathbb{Z} \to 0$ and then to the class of $r_{\mathbb{Z}} \circ s_F$, where s_F is a right inverse of $W_0E \otimes \mathbb{C} \xrightarrow{\pi}\!\!\!\to \mathbb{C}$, preserving the Hodge filtration, and $r_{\mathbb{Z}}$ is a left inverse of $W_0H \xrightarrow{i} W_0E$ (cf. [Ca] lemma 4). The first map in 9.2.3 is an isomorphism since $\mathrm{Hom}_{MH}(\mathbb{Z}, H/W_0H) = 0 = \mathrm{Ext}^1_{MH}(\mathbb{Z}, H/W_0H)$: any extension

$$0 \longrightarrow H' \longrightarrow E' \longrightarrow \mathbb{Z} \longrightarrow 0 \quad , \quad W_0H' = 0 ,$$

is split by W_0E'. The second map in 9.2.3 is an isomorphism by the same arguments as in [Ca].

For the second claim compare loc. cit. 2d ; a complete proof can be obtained by combining [EV] 7.11, where the Abel-Jacobi map is defined via a method of El Zein and Zucker, which amounts to the composition of 9.0.3 and 9.2.3, and [J2] § 1, where the agreement with the classical Abel-Jacobi map is shown.

<u>9.3. Remarks</u> a) For mixed Hodge structures A and B one has an exact sequence

$$0 \longrightarrow \mathrm{Ext}^1_{MH}(\mathbb{Z}, \underline{\mathrm{Hom}}(A,B)) \xrightarrow{\alpha} \mathrm{Ext}^1_{MH}(A,B) \xrightarrow{\beta} \mathrm{Ext}^1_{\mathbb{Z}}(A,B) \longrightarrow 0 ,$$

and an isomorphism

$$\mathrm{Ext}^1_{MH}(A,B) \otimes \mathbb{Q} \xrightarrow{\sim} \mathrm{Ext}^1_{\mathbb{Q}-MH}(A \otimes \mathbb{Q}, B \otimes \mathbb{Q}) .$$

The morphism β is obvious, and α maps an extension

$$0 \longrightarrow \underline{\mathrm{Hom}}(A,B) \longrightarrow E \longrightarrow \mathbb{Z} \longrightarrow 0$$

to the push-out of

$$0 \longrightarrow \underline{\mathrm{Hom}}(A,B) \otimes A \longrightarrow E \otimes A \longrightarrow A \longrightarrow 0$$

via the evaluation map $\underline{\mathrm{Hom}}(A,B) \otimes A \longrightarrow B$.

b) One has $\mathrm{Ext}^i_{MH}(A,B) = 0$ for $i \geq 2$. This is proved in [Bei 2]; it also follows from the right-exactness of $\mathrm{Ext}^1_{MH}(A,B)$, which is clear from the explicit description above.

We now turn to the ℓ-adic realizations. First let X be smooth; then we obtain a homomorphism

$$\begin{aligned} z^j(X)_0 &\longrightarrow \mathrm{Ext}^1_{G_k}(\mathbb{Z}_\ell, H^{2j-1}(\bar{X}, \mathbb{Z}_\ell(j))) \\ &= H^1_{cont}(G_k, H^{2j-1}(\bar{X}, \mathbb{Z}_\ell(j))) \end{aligned} \tag{9.4.1}$$

<u>9.4. Lemma</u> The map above is induced by the cycle map

$$z^j(X) \longrightarrow H^{2j}_{cont}(X, \mathbb{Z}_\ell(j))$$

and the Hochschild-Serre spectral sequence

$$E_2^{p,q} = H^p_{cont}(G_k, H^q(\bar{X}, \mathbb{Z}_\ell(j))) \Rightarrow H^{p+q}_{cont}(X, \mathbb{Z}_\ell(j)) \tag{9.4.1}$$

(see [J1] 3.5 b)), i.e., by the edge morphism

$$\mathrm{Ker}(H^{2j}_{cont}(X, \mathbb{Z}_\ell(j)) \xrightarrow{res} H^{2j}(\bar{X}, \mathbb{Z}_\ell(j)) \longrightarrow H^1_{cont}(G_k, H^{2j-1}(\bar{X}, \mathbb{Z}_\ell(j)))$$

(cf. the discussion in [J1] 6.15).

This will follow from the next general fact.

9.5. Lemma Let $\mathcal{A}$ be an abelian category with enough injectives and let

$$0 \longrightarrow A^\cdot \xrightarrow{\alpha} B^\cdot \xrightarrow{\beta} C^\cdot \longrightarrow 0$$

be an exact sequence of bounded below complexes in $\mathcal{A}$. Let $f : \mathcal{A} \longrightarrow \mathcal{B}$ be a left exact functor into another abelian category and denote by F^i the decreasing filtration induced on the limit terms by the hypercohomology spectral sequence

$$(9.5.1) \qquad E_2^{p,q} = R^p f \mathcal{H}^q(K^\cdot) \Rightarrow \mathbb{R}^{p+q} f K^\cdot$$

for a bounded below complex $K^\cdot$ in $\mathcal{A}$ (here $\mathcal{H}^q(K^\cdot)$ is the homology of $K^\cdot$ at the q-th place).

Fix an $n \in \mathbb{Z}$ and let $X = \operatorname{Ker} \partial$ and $Y = \operatorname{Im} \partial$ in the long exact sequence

$$\begin{array}{ccccccccc} \cdots \longrightarrow & \mathcal{H}^{n-1}(B^\cdot) & \longrightarrow & \mathcal{H}^{n-1}(C^\cdot) & \xrightarrow{\partial} & \mathcal{H}^n(A^\cdot) & \longrightarrow & \mathcal{H}^n(B^\cdot) & \longrightarrow \cdots \\ & & {\scriptstyle\pi}\searrow \quad \nearrow & & \searrow \quad \nearrow & & & & \\ (9.5.2) & & X & & Y & & & & \end{array}$$

Let $Y' = \rho_0^{-1}(fY) = \alpha_*^{-1}(F^1\, \mathbb{R}^n f B^\cdot)$ in the commutative diagram

$$\begin{array}{ccccccc}
0 \longrightarrow & F^1\,\mathbb{R}^n fB^{\cdot} & \longrightarrow & \mathbb{R}^n fB^{\cdot} & \longrightarrow & fH^n(B^{\cdot}) \\
& \uparrow & \nwarrow \tau & \uparrow \alpha_* & & \uparrow \\
(9.5.3) \qquad 0 \longrightarrow & F^1\,\mathbb{R}^n fA^{\cdot} & \longrightarrow & \mathbb{R}^n fA^{\cdot} & \xrightarrow{\rho_0} & fH^n(A^{\cdot}) \\
& & & \cup\!\mid & & \uparrow \\
& & & Y' & \longrightarrow & fY
\end{array}$$

with exact rows. As indicated, let τ be the arrow induced by α_* , and let $\rho_1 : F^1\,\mathbb{R}^n fB^{\cdot} \longrightarrow R^1 fH^{n-1}(B^{\cdot})$ be the edge morphism from the spectral sequence 9.5.1.

Then the diagram

$$(9.5.4) \qquad \begin{array}{ccccc}
& & fY & \xrightarrow{\delta} & R^1 fX \\
& \nearrow & & & \uparrow R^1 f(\pi) \\
Y' & & & & \\
& \searrow^{\tau} & & & \\
& & F^1\,\mathbb{R}^n fB^{\cdot} & \xrightarrow{\rho_1} & R^1 fH^{n-1}(B^{\cdot})
\end{array}$$

commutes, where δ is the connecting morphism for the exact sequence

$$(9.5.5) \qquad 0 \longrightarrow X \longrightarrow H^{n-1}(C^{\cdot}) \longrightarrow Y \longrightarrow 0 \ .$$

<u>Proof</u> We may assume that $A^{\cdot}$, $B^{\cdot}$ and $C^{\cdot}$ have injective components. Consider the commutative diagram with exact rows

$$
(9.5.6)\qquad
\begin{array}{ccccccccc}
0 & \longrightarrow & A^{n+1} & \longrightarrow & B^{n+1} & \longrightarrow & C^{n+1} & \longrightarrow & 0 \\
 & & \uparrow{\scriptstyle d_A^n} & & \uparrow{\scriptstyle d_B^n} & & \uparrow{\scriptstyle d_C^n} & & \\
0 & \longrightarrow & A^{n} & \longrightarrow & B^{n} & \longrightarrow & C^{n} & \longrightarrow & 0 \\
 & & \uparrow{\scriptstyle d_A^{n-1}} & & \uparrow{\scriptstyle d_B^{n-1}} & & \uparrow{\scriptstyle d_C^{n-1}} & & \\
0 & \longrightarrow & A^{n-1} & \longrightarrow & B^{n-1} & \longrightarrow & C^{n-1} & \longrightarrow & 0
\end{array}
$$

This induces a commutative diagram with exact rows

$$
(9.5.7)\qquad
\begin{array}{ccccccccc}
0 \to & H^{n-1}(B^\cdot) & \to & B^{n-1}/\mathrm{Im}\ d_B^{n-2} & \xrightarrow{d_B^{n-1}} & \mathrm{Im}\ d_B^{n-1} & \to 0 \\
 & \Vert & & \cup\!\mid & & \cup\!\mid & \\
0 \to & \mathrm{Ker}\ d_B^{n-1}/\mathrm{Im}\ d_B^{n-2} & \to & (d_B^{n-1})^{-1}\mathrm{Im}\ \alpha/\mathrm{Im}\ d_B^{n-2} & \to & \mathrm{Im}\ \alpha \cap \mathrm{Im}\ d_B^{n-1} & \to 0 \\
 & \pi\big\downarrow & & \beta\big\downarrow & & \big\downarrow & \\
0 \to & X & \to & \mathrm{Ker}\ d_c^{n-1}/\mathrm{Im}\ d_c^{n-2} & & Y & \to 0
\end{array}
$$

We obtain a commutative diagram for the connecting morphisms δ

$$
(9.5.8)\qquad
\begin{array}{ccc}
f(\mathrm{Im}\ d_B^{n-1}) & \xrightarrow{\delta} & R^1 fH^{n-1}(B^\cdot) \\
\uparrow & & \Vert \\
f(\mathrm{Im}\ \alpha \cap \mathrm{Im}\ d_B^{n-1}) & \xrightarrow{\delta} & R^1 fH^{n-1}(B^\cdot) \\
\big\downarrow & & \big\downarrow {\scriptstyle R^1 f(\pi)} \\
fY & \xrightarrow{\delta} & R^1 fX
\end{array}
$$

By the assumption on $B^\cdot$ we get a commutative exact diagram

$$
(9.5.9)\qquad
\begin{array}{ccccc}
0 \to F^1\,\mathbb{R}^n fB^\cdot & \to & \mathbb{R}^n fB^\cdot & \to & fH^n(B^\cdot) \\
\uparrow\wr & & \| & & \| \\
0 \to f\,\mathrm{Im}\,d_B^{n-1}/\mathrm{Im}\,f(d_B^{n-1}) & \to & \mathrm{Ker}\,f(d_B^n)/\mathrm{Im}\,f(d_B^{n-1}) & \to & f(\mathrm{Ker}\,d_B^n/\mathrm{Im}\,d_B^{n-1})
\end{array}
$$

and a similar one for $A^\cdot$, giving a canonical isomorphism

$$Y' = f(\alpha)^{-1}(f\ \mathrm{Im}\ d_B^{n-1})/\mathrm{Im}\ f(d_A^{n-1}) \cong f(\mathrm{Im}\ \alpha \cap \mathrm{Im}\ d_B^{n-1})/f(\alpha)(\mathrm{Im}\ f(d_A^{n-1})).$$

So the diagram 9.5.4 is obtained from 9.5.8 via factorization, once we have seen that

$$
\begin{array}{ccc}
F^1\,\mathbb{R}^n fB^\cdot & \xrightarrow{\ \rho_1\ } & R^1 fH^{n-1}(B^\cdot) \\
\Big\uparrow & \nearrow_{\delta} & \\
f\ \mathrm{Im}\ d_B^{n-1} & &
\end{array}
$$

is commutative. Let τ_n be the canonical increasing filtration of $B^\cdot$, then ρ_1 is given by the isomorphism

$$F^1\,\mathbb{R}^n fB^\cdot = \mathrm{Im}(\,\mathbb{R}^n f\tau_{n-1}B^\cdot \longrightarrow \mathbb{R}^n fB^\cdot) \cong \mathbb{R}^n f\tau_{n-1}B^\cdot$$

(note that $\mathbb{R}^{n-1}f(B^\cdot/\tau_{n-1}B^\cdot) = 0$) and the morphism

$$\mathbb{R}^n f\tau_{n-1}B^\cdot \longrightarrow \mathbb{R}^n f(H^{n-1}(B^\cdot)[-n+1]) = R^1 fH^{n-1}(B^\cdot)$$

induced by

$$
\begin{array}{llccccccc}
\tau_{n-1}B^{\cdot} & : \ \ldots & B^{n-3} & \rightarrow & B^{n-2} & \longrightarrow & \operatorname{Ker} d_B^{n-1} & \rightarrow & 0 \ \ldots \\
\big\downarrow \text{can} & & \big\downarrow & & \big\downarrow & & \big\downarrow & & \big\downarrow \\
\mathcal{H}^{n-1}(B^{\cdot})[-n+1] & : \ \ldots & 0 & \rightarrow & 0 & \longrightarrow & \mathcal{H}^{n-1}(B^{\cdot}) & \rightarrow & 0 \ \ldots
\end{array}
$$

Diagram (9.5.10) is now obtained by applying $\mathbb{R}^n f$ to the commutative diagram in the derived category

$$
\begin{array}{ccc}
\tau_{n-1}B^{\cdot} & \xrightarrow{\text{can}} & \mathcal{H}^{n-1}(B^{\cdot})[-n+1] \\
\uparrow & \nearrow & \\
\operatorname{Im} d_B^{n-1}[-n] & &
\end{array}
$$

obtained from the commutative diagram

$$
\begin{array}{lccccccc}
\operatorname{Im} d_B^{n-1}[-n] & = & \ldots 0 & \rightarrow & 0 & \rightarrow & \operatorname{Im} d_B^{n-1} & \rightarrow 0 \ \ldots \\
\big\downarrow & & \big\downarrow & & \big\downarrow & & \| & \big\downarrow \\
\tau_{n-1}B^{\cdot} & \xrightarrow{\sim} & \ldots B^{n-2} & \rightarrow & B^{n-1} & \rightarrow & \operatorname{Im} d_B^{n-1} & \rightarrow 0 \ \ldots \\
\big\downarrow \text{can} & & \big\downarrow & & \big\downarrow & & \| & \big\downarrow \\
\mathcal{H}^{n-1}(B^{\cdot})[-n+1] & \xrightarrow{\sim} & \ldots 0 & \rightarrow & B^{n-1}/\operatorname{Ker} d_B^{n-1} & \rightarrow & \operatorname{Im} d_B^{n-1} & \rightarrow 0 \ \ldots
\end{array}
$$

where $\xrightarrow{\sim}$ stands for a quasiisomorphism. q.e.d.

9.6. To obtain 9.4, we apply 9.5 to the exact triangle

$$(9.6.1) \qquad Rg_*\nu_*R\nu^! \mathbb{Z}_\ell(j) \longrightarrow Rg_* \mathbb{Z}_\ell(j) \longrightarrow Rg_*R\mu_*\mu^* \mathbb{Z}_\ell(j) \longrightarrow$$

associated to the diagram

$$\begin{array}{ccccc} Z & \overset{\nu}{\hookrightarrow} & X & \overset{\mu}{\hookleftarrow} & U = X \setminus Z \\ & {\scriptstyle g_1}\searrow & \downarrow{\scriptstyle g} & \swarrow{\scriptstyle g_2} & \\ & & \operatorname{Spec} k & & \end{array} .$$

If $\mathbb{Z}_\ell(j) \hookrightarrow I^\cdot$ is a resolution by injective pro-ℓ-systems of étale sheaves on X (i.e., injective objects in $S(X_{et})^{\mathbb{Z}_\ell}$, see [J1] 6.9), 9.6.1 is represented by the short exact sequence of complexes

$$(9.6.2) \qquad 0 \longrightarrow g_*\nu_*\nu^! I^\cdot \longrightarrow g_*I^\cdot \longrightarrow g_*\mu_*\mu^* I^\cdot \longrightarrow 0$$

whose components are injective objects in $S(\operatorname{Spec} k_{et})^{\mathbb{Z}_\ell}$. Identifying noetherian ℓ-adic sheaves on $\operatorname{Spec} k_{et}$ with objects in $\operatorname{Rep}_c(G_k, \mathbb{Z}_\ell)$, the long exact sequence 9.5.2 then is

$$\begin{array}{l} \ldots\ H^{n-1}(\bar{X}, \mathbb{Z}_\ell(j)) \longrightarrow H^{n-1}(\bar{U}, \mathbb{Z}_\ell(j)) \longrightarrow H^n_{\bar{Z}}(\bar{X}, \mathbb{Z}_\ell(j)) \longrightarrow H^n(\bar{X}, \mathbb{Z}_\ell(j)) \longrightarrow \ldots , \\ \qquad\qquad \searrow\!\!\searrow \ \nearrow \qquad\qquad\qquad\quad \searrow\!\!\searrow \ \nearrow \\ \qquad\qquad\quad R \qquad\qquad\qquad\qquad\qquad\quad S \end{array}$$

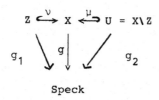

while the continuous cohomology groups over k, $H_{cont,Z}(X, \mathbb{Z}_\ell(j))$, $H_{cont}(X, \mathbb{Z}_\ell(j))$ and $H^\cdot_{cont}(U, \mathbb{Z}_\ell(j))$, and their relative cohomology sequence, is obtained by applying $RH^0_{cont}(G_k,-)$ to the triangle 9.6.1 (i.e., $f = \varprojlim_n H^0(G_k,-)$ to the sequence 9.6.2) and taking the associated long exact cohomology sequence.

Now by definition the pull-back of the extension

$$0 \longrightarrow R \longrightarrow H^{n-1}(\bar{U}, \mathbb{Z}_\ell(j)) \longrightarrow S \longrightarrow 0$$

via an element $z \in S^{G_k} = \mathrm{Hom}_{G_k}(\mathbb{Z}_\ell, S)$ corresponds to the image of z under the connecting morphism

$$S^{G_k} \longrightarrow H^1_{cont}(G_k,R) = \mathrm{Ext}^1_{G_k}(\mathbb{Z}_\ell,R) \ ,$$

and by 9.5 this image is obtained via the Hochschild-Serre spectral sequence (which by definition is the hypercohomology spectral sequence for $Rg_* \mathbb{Z}_\ell(j)$) in the claimed way.

<u>9.7. Remarks</u> a) Similar results hold of course for finite coefficients $\mathbb{Z}/\ell^r(j)$ and usual étale cohomology, and the result above can be deduced from this since one has

$$(9.7.1) \quad H^1_{cont}(G_k,H^n(\bar{X}, \mathbb{Z}_\ell(j)) = \varprojlim_r H^1(G_k,H^n(\bar{X}, \mathbb{Z}/\ell^r(j)))$$

(because $\varprojlim_r{}^1 H^0(G_k,H^n(\bar{X}, \mathbb{Z}/\ell^r(j)) = 0$, cf. [J1] 2.1). If $H^{i-1}(G_k,H^n(\bar{X}, \mathbb{Z}/\ell^r(j)))$ is infinite (e.g., for number fields), this formula becomes false for $H^i_{cont}(G_k,-)$, $i > 1$, and there is no Hochschild-Serre spectral sequence for the "naive" ℓ-adic cohomology $H^i(X, \mathbb{Z}_\ell(j)) = \varprojlim_r H^i(X, \mathbb{Z}/\ell^r(j))$ (by the non-exactness of the inverse limit). For the same reason, one has to be careful with passing to the inverse limit in the various exact sequences of 9.5, so I have preferred to work with 9.4.1.

b) For possibly singular X a similar statement as in 9.4 holds for homology, in terms of the Abel-Jacobi map

$$(9.7.2)\qquad Z_i(X)_0 \longrightarrow H^1_{cont}(G_k, H_{2i+1}(\bar{X}, \mathbb{Z}_\ell(i)))$$

and the "Hochschild-Serre spectral sequence"

$$(9.7.3)\qquad E_2^{p,q} = H^p_{cont}(G_k, H_{-q}(\bar{X}, \mathbb{Z}_\ell(i))) \Rightarrow H^{cont}_{-p-q}(X, \mathbb{Z}_\ell(i)) .$$

For finite coefficients $\mathbb{Z}/\ell^r(i)$, étale homology $H_a(X, \mathbb{Z}/\ell^r(i))$ is defined as hypercohomology $H^{-a}(X, Rg^! \mathbb{Z}/\ell^r(-i))$ of the twisted dualizing complex defined by duality theory (cf. 6.7). Everything is proved as above, replacing in 9.6 the sheaf $\mathbb{Z}_\ell(j)$ by $Rg^! \mathbb{Z}/\ell^r(-i)$; the spectral sequence 9.7.3 (for finite coefficients) is by definition the hypercohomology spectral sequence for $Rg_* Rg^! \mathbb{Z}/\ell^r(-i)$ and (the derivatives of) the functor $H^0(G_k, -)$, and one uses the relations $Rg_* \nu_* R\nu^! Rg^! = R(g_1)_* Rg_1^!$ and $Rg_* R\mu_* \mu^* Rg^! = R(g_2)_* Rg_2^!$.

For $\mathbb{Z}_\ell$-coefficients one uses a certain complex $Rg^! \mathbb{Z}_\ell(-i)$ in $D^b(S(X_{et})^{\mathbb{Z}_\ell})$, whose components in $D^b(S(X_{et}, \mathbb{Z}/\ell^r))$ are the complexes $Rg^! \mathbb{Z}/\ell^r(-i)$ above, and its continuous étale hypercohomology. For the existence of $Rg^! \mathbb{Z}_\ell(-i)$ and a cycle map

$$cl : Z_i(X) \longrightarrow H^{cont}_{2i}(X, \mathbb{Z}_\ell(i)) = H^{-2i}_{cont}(X, Rf^! \mathbb{Z}_\ell(-i))$$

we refer the reader to a future paper [J5].

c) The Hodge theory also fits into the scheme of 9.5. Namely, Beilinson [Bei 2] has shown that to each variety X over $\mathbb{C}$ and $j \in \mathbb{Z}$ there is associated a complex

$$\underline{R\Gamma}(X, \mathbb{Z}(j)) \in D^b(\mathcal{MH})$$

of mixed Hodge structures such that its n-th homology $\mathcal{H}^n$ is just the singular cohomology $H^n(X, \mathbb{Z})(j)$ with its mixed Hodge structures and such that the n-th homology of

(9.7.4) $$R\Gamma_{\mathcal{H}}\underline{R\Gamma}(X, \mathbb{Z}(j)) := R\mathrm{Hom}_{D^b(\mathcal{MH})}(\mathbb{Z}(o), \underline{R\Gamma}(X, \mathbb{Z}(j)))$$

coincides with the Deligne cohomology $H^n_{\mathcal{D}}(X, \mathbb{Z}(j))$ for $n \leq 2j$. Moreover in this case the exact sequences

(9.7.5) $$0 \to R^1\Gamma_{\mathcal{H}}H^{n-1}(X, \mathbb{Z}(j)) \to H^n_{\mathcal{D}}(X, \mathbb{Z}(j)) \xrightarrow{\varepsilon_{\mathbb{Z}}} \Gamma_{\mathcal{H}}H^n(X, \mathbb{Z}(j)) \to 0$$

coming from the hypercohomology spectral sequence and the vanishing of $R^i\Gamma_{\mathcal{H}}$ for $i \geq 2$ (cf. 9.3 b)) coincide with the exact sequences coming from the usual definition of Deligne cohomology ([Bei 1] 1.6 (*) and [EV] 2.10 a)), via the isomorphisms 9.2.1 and the relation

(9.7.6) $$\Gamma_{\mathcal{H}}H = \mathrm{Hom}_{\mathcal{MH}}(\mathbb{Z}(0), H) \cong W_0H \cap F^0H \otimes \mathbb{C}$$

for a mixed Hodge structure H.

There are versions with support, for homology etc., and the Hodge version of 9.1.1 is induced by an exact triangle

$$\underline{R\Gamma}_Z(X, \mathbb{Z}(j)) \longrightarrow \underline{R\Gamma}(X, \mathbb{Z}(j)) \longrightarrow \underline{R\Gamma}(U, \mathbb{Z}(j)) \longrightarrow$$

in $D^b(\mathcal{MH})$ and by the cycle class of z in $H^{2j}_{\mathcal{D},Z}(X, \mathbb{Z}(j)) = \Gamma_{\mathcal{H}}H^{2j}_Z(X, \mathbb{Z}(j))$. The latter defines a cycle map

$$c\ell_{\mathcal{D}} : Z^j(X) \longrightarrow H^{2j}_{\mathcal{D}}(X, \mathbb{Z}(j)) \quad ,$$

whose composition with $\varepsilon_{\mathbb{Z}}$ is the cycle map into singular cohomology discussed before. By 9.5 the Abel-Jacobi map fits into a commutative diagram

$$\begin{array}{ccccccccc}
0 \to & Z^j(X)_0 & \to & Z^j(X) & \to & Z^j(X)/Z^j(X)_0 & \to 0 \\
& \downarrow c\ell' & & \downarrow c\ell & & \downarrow c\ell & \\
0 \to & R^1\Gamma_{\mathcal{H}} H^{2j-1}(X, \mathbb{Z}(j)) & \to & H^{2j}_{\mathcal{D}}(X, \mathbb{Z}(j)) & \to & \Gamma_{\mathcal{H}} H^{2j}(X, \mathbb{Z}(j)) & \to 0 \quad ,
\end{array}$$

compare [Bei 1] 1.9, [Bei 2] 5.2. If X is no longer smooth, similar statements hold for homology (loc. cit. and [J2]).

We get the analogy (compare [Bei 4] 3.0):

<u>geometric theories</u>

$H^i(\bar{X}, \mathbb{Z}_\ell(j))$ with G_k-action $\longleftrightarrow$ $H^i(X(\mathbb{C}), \mathbb{Z}(j))$ with Hodge structure

<u>absolute theories</u>

$H^i_{cont}(X, \mathbb{Z}_\ell(j))$ $\longleftrightarrow$ $H^i_{\mathcal{D}}(X, \mathbb{Z}(j))$

<u>connection</u>

Hochschild-Serre spectral sequence 9.4.1	$\longleftrightarrow$	short exact sequence 9.7.5

with the difference that $R^i\Gamma_{G_k} = H^i_{cont}(G_k,-)$ in general does not vanish for $i \geq 2$, e.g., if k is a global field. We shall return to this point later on (e.g., in 11.4 c) or 12.19).

Now we want to show the connection between extension classes and the conjectures of the previous chapters. We go back to the general setting of an F-linear twisted Poincaré duality theory with weights (F a field of characteristic zero) and transformations

$$r = ch_{i,j} : H^i_{M,Z}(X,\mathbb{Q}(j)) \longrightarrow \Gamma H^i_Z(X,j)$$

$$r' = \tau_{a,b} : H^M_a(X,\mathbb{Q}(b)) \longrightarrow \Gamma H_a(X,b)$$

as in 8.4, and we suppose that these are compatible with the cycle map as in 8.4.2.

<u>9.8. Lemma</u> Let X be smooth and proper of pure dimension d and let $Z \subset X$ be of codimension $\geq j$, then there is a commutative diagram

$$(9.8.1)\qquad \begin{array}{ccccc}
0 \to \Gamma H^{2j-1}(U,j) & \to & \Gamma H^{2j}_Z(X,j)_0 & \xrightarrow{\delta} & R^1\Gamma H^{2j-1}(X,j) \\
 & & \uparrow{\scriptstyle c\ell} & (3) & \uparrow{\scriptstyle c\ell'} \\
\uparrow{\scriptstyle ch_{2j-1,j}} & (1) & CH^j_Z(X)_0 \otimes F & \to & CH^j(X)_0 \otimes F \\
 & & \uparrow{\scriptstyle ch_{2j,j}}\,\wr & (2) & \wr\uparrow{\scriptstyle ch_{2j,j}} \\
H^{2j-1}_M(U,\mathbb{Q}(j)) \otimes F & \to & H^{2j}_{M,Z}(X,\mathbb{Q}(j))_0 \otimes F & \to & H^{2j}_M(X,\mathbb{Q}(j))_0 \otimes F ,
\end{array}$$

where $c\ell'$ is induced by the Abel-Jacobi map 9.0.3,

$A_0 = \mathrm{Ker}(A \longrightarrow \Gamma H^{2j}(X,j))$ for the various groups A in the diagram, and the top exact sequence is part of the long exact $\mathrm{Ext}_T(1,-)$-sequence associated to the short exact sequence

$$0 \longrightarrow H^{2j-1}(X,j) \longrightarrow H^{2j-1}(U,j) \longrightarrow H^{2j}_Z(X,j)_0 \longrightarrow 0 .$$

(Note that $\Gamma H^{2j-1}(X,j) = 0$ since $H^{2j-1}(X,j)$ is pure of weight -1) .

<u>Proof</u> The commutativity of (1) follows from the compatibility of ch with the relative cohomology sequences, since $c\ell \circ ch_{2j,j} = ch_{2j,j}$ by assumption. Similarly, (2) is commutative, and $\delta \circ c\ell$ by definition is the Abel-Jacobi map 9.0.3 on $CH^j_Z(X)_0 \otimes F = CH_{d-j}(Z)_0 \otimes F = (\bigoplus_{x \in X^{(j)} \cap Z} \mathbb{Z})_0 \otimes F$ (explicit description of the connecting morphism δ in the long exact Ext-sequence, compare 9.6 above). Hence (3) is commutative once we have seen that the Abel-Jacobi map factorizes through rational equivalence. But by the bottom exact sequence and (2) the image of $H^{2j-1}_{\mathcal{M}}(U,\mathbb{Q}(j)) \otimes F$ in $(\bigoplus_{x \in X^{(j)} \cap Z} \mathbb{Z}) \otimes F$ is the subgroup generated by cycles linearly equivalent to zero, and by (2) it is mapped to zero under $\delta \circ c\ell$. Since Z is arbitrary, this shows the factorization for all cycles in $Z^j(X)_0$.

<u>9.9. Lemma</u> The middle vertical map in 9.8.1 is an isomorphism under the assumption p) and the following one (which is satisfied for the examples 8.3).

q) For irreducible Z of dimension e the fundamental class $\eta_Z : 1 \longrightarrow H_{2e}(Z,e)$ is an isomorphism, and $\Gamma(1) = F$.

<u>Proof</u> We have $\Gamma H_Z^{2j}(X,j) \cong \Gamma H_{2(d-j)}(Z,d-j) \cong \bigoplus_{x\in Z^{(0)}} F$ $\xleftarrow[\sim]{c\ell} (\bigoplus_{x\in X^{(j)}\cap Z} \mathbb{Z}) \otimes F = CH_Z^j(X) \otimes F$ (by using 6.1 f) and p) one reduces to the case that Z is disjoint union of its irreducible components).

<u>9.10. Corollary</u> Assume p) and q). Let X be smooth and proper, let $Z \subset X$ be of pure codimension j, and let $U = X - Z$. Then

$$ch_{2j-1,j} : H_{\mathcal{M}}^{2j-1}(U,\mathbb{Q}(j)) \otimes F \longrightarrow \Gamma H^{2j-1}(U,j)$$

is surjective if and only if the Abel-Jacobi map

$$c\ell_j^! : CH^j(X)_0 \otimes F \longrightarrow R^1\Gamma H^{2j-1}(X,j)$$

is injective on the subgroup generated by the cycles with support on Z.

<u>Proof</u> This follows immediately from diagram 9.8.1 and the bijectivity of the middle vertical map (without restriction X is connected).

<u>9.11. Corollary</u> For smooth varieties U over $\mathbb{C}$ the Chern character

$$ch_{2j-1,j} : H_{\mathcal{M}}^{2j-1}(U,\mathbb{Q}(j)) \longrightarrow \Gamma_H(H^{2j-1}(U,\mathbb{Q}(j)))$$

is in general not surjective for $j \geq 2$ (This disproves a conjecture of Beilinson, [Bei 2] Conjecture 6).

<u>Proof</u> The complex Abel-Jacobi map is known to be injective for

$j = 1$ (this gives a new proof for 5.13, at least $\mathbb{Q}$-rationally), but Mumford has shown that it is in general not injective for codimension $j \geq 2$, not even up to torsion ([Mu]) : For a smooth and proper variety of dimension d over a field k let $\mathrm{Alb}(X)$ be the Albanese variety and let

$$(9.11.1) \qquad T(X) = \mathrm{Ker}(A_0(X) \longrightarrow \mathrm{Alb}(X)(k)) ,$$

where $A_0(X) \subseteq CH_0(X)$ is the subgroup of cycles of degree zero. For $k = \mathbb{C}$ one has $\mathrm{Alb}(X)(\mathbb{C}) \xrightarrow{\sim} J^d_{\mathbb{C}}(X)$, where

$$J^j_{\mathbb{C}}(X) = H^{2j-1}(X,\mathbb{C})/H^{2j-1}(X, \mathbb{Z}(j)) + F^jH^{2j-1}(X,\mathbb{C})$$

is the j-th intermediate Jacobian. Hence $T(X)$ is the kernel of the Abel-Jacobi map, and for $d = 2$ Mumford shows

$$(9.11.2) \qquad p_g(X) > 0 \Rightarrow T(X) \otimes \mathbb{Q} \neq 0$$

(in fact, $T(X)$ is huge in this case), where $p_g(X) = \dim_{\mathbb{C}} H^2(X,\mathcal{O}_X)$ is the geometric genus of the surface X. This was generalized by Roitman [Ro 1] to the following result for arbitrary d and arbitrary uncountable fields k of characteristic zero:

$$(9.11.3) \quad H^0(X,\Omega^p_X) \neq 0 \quad \text{for some} \quad p \geq 2 \Rightarrow T(X) \otimes \mathbb{Q} \neq 0 .$$

Now why should conjecture 5.20 be true? The answer is the following. Mumford's counterexample involves cycles which are defined over fields of higher transcendence degree ("generic cycles"), and it is expected that the situation is completely different over number fields. Namely, Bloch and Beilinson have independently conjectured the following ([Bei 4] 5.6):

9.12. Conjecture For a smooth, proper variety X over $\mathbb{Q}$ the complex Abel-Jacobi map

$$CH^j(X)_0 \longrightarrow J^j_{\mathbb{C}}(X \times_{\mathbb{Q}} \mathbb{C})$$

is injective up to torsion (here $CH^j(X)$ is the Chow group over $\mathbb{Q}$).

Since in the situation of 9.10 U is defined over $\mathbb{Q}$ if X and Z are defined over $\mathbb{Q}$, conjecture 9.12 is implied by conjecture 5.20 for $i = 2j-1$, for a converse see 9.18 below. Actually, Bloch and Beilinson only consider the subgroups of cycles algebraically equivalent to zero, but from our approach there should be no difference. Also, by 5.21 it is the same to consider arbitrary number fields instead of $\mathbb{Q}$.

Let us now consider the ℓ-adic case. Bloch has remarked that in Mumford's situation Hodge theory and Lefschetz' theorem on 1-cycles give the equivalence

$$p_g(X) > 0 \iff H^2(X,\mathbb{C}) \neq H^{1,1} \iff NS(X) \otimes \mathbb{C} \neq H^2(X,\mathbb{C}) \quad ,$$

where $NS(X)$ is the Néron-Severi group of X, and he has proved the following generalization of Mumford's theorem to arbitrary fields k.

9.13 Theorem ([Bl 1] 1.24) Let X be a connected, smooth, proper surface over a field k, with function field K, and let $\ell \neq \operatorname{char} k$ be a prime. If $NS(\bar{X}) \otimes \mathbb{Q}_\ell \neq H^2_{ét}(\bar{X},\mathbb{Q}_\ell(1))$, then $T(X \times_k K) \otimes \mathbb{Q} \neq 0$.

This is not stated in this form in loc. cit., but follows from the proof. We now investigate the relations between $T(X)$ and the ℓ-adic Abel-Jacobi map.

<u>9.14. Lemma</u> Let X be smooth and proper of dimension d, then $T(X)$ lies in the kernel of the ℓ-adic Abel-Jacobi map

$$A_0(X) = CH_0(X)_0 \xrightarrow{c\ell'} H^1_{cont}(G_k, H^{2d-1}(\bar{X}, \mathbb{Z}_\ell(d))) ,$$

$\ell \neq char(k)$. If k is finitely generated, then the natural map

$$Alb(X)(k) \otimes \mathbb{Z}_\ell \longrightarrow H^1_{cont}(G_k, H^{2d-2}(\bar{X}, \mathbb{Z}_\ell(d)))$$

is injective, so that $T(X) \otimes \mathbb{Z}_\ell = Ker\ c\ell' \otimes \mathbb{Z}_\ell$

<u>Proof</u> First observe that $A_0(X) = CH^d(X)_0$ for the homological equivalence given by the ℓ-adic cohomology, by the commutative diagram

$$\begin{array}{ccc} & \overset{cl}{\nearrow} & H^{2d}(\bar{X}, \mathbb{Z}_\ell(d)) \\ CH^d(X) & & \wr\downarrow tr \\ & \underset{deg}{\searrow} & \mathbb{Z} \subset \mathbb{Z}_\ell , \end{array}$$

where tr is the canonical trace map. It is well-known that the groups $A_0(X \times_k \bar{k})$ and $Alb(X)(\bar{k})$ are divisible, and by a theorem of Roitman ([Ro 2] 3.1) the first vertical map in the commutative diagram

(9.14.1)
$$\begin{array}{ccccccccc} 0 & \longrightarrow & {}_{\ell^n}A_0(\bar{X}) & \longrightarrow & A_0(\bar{X}) & \xrightarrow{\ell^n} & A_0(\bar{X}) & \longrightarrow & 0 \\ & & \downarrow & & \downarrow & & \downarrow & & \\ 0 & \longrightarrow & {}_{\ell^n}Alb(X)(\bar{k}) & \longrightarrow & Alb(X)(\bar{k}) & \xrightarrow{\ell^n} & Alb(X)(\bar{k}) & \longrightarrow & 0 \end{array}$$

is an isomorphism. Since $Alb(X)$ is the dual abelian variety to $\underline{Pic}^0(X)$, one has isomorphisms

$$ {}_{\ell^n}Alb(X)(\bar{k}) \cong Hom(H^1(\bar{X},\mu_{\ell^n}),\mu_{\ell^n}) \cong H^{2d-1}(\bar{X},\mathbb{Z}/\ell^n(d)) $$

by Poincaré duality, and one easily shows (see [J5]) that the map

$$ \begin{array}{ccccc} A_0(X) \longrightarrow Alb(X)(k) = Alb(X)(\bar{k})^{G_k} & \longrightarrow & \varprojlim_n H^1(G_k, H^{2d-1}(X,\mathbb{Z}/\ell^n(d))) \\ & & \| \\ & & H^1_{cont}(G_k, H^{2d-1}(\bar{X},\mathbb{Z}_\ell(d))) \end{array} $$

induced by the Kummer sequences 9.14.1 coincides with cl'. This shows the statement for $T(X)$ and gives an injection

$$ Alb(X)(k)^\wedge := \varprojlim_n Alb(X)(k)/\ell^n \hookrightarrow H^1_{cont}(G_k, H^{2d-1}(\bar{X},\mathbb{Z}_\ell(d))) . $$

If k is finitely generated, then $Alb(X)(k)$ is a finitely generated abelian group by the generalized Mordell-Weil theorem, and hence $Alb(X)(k) \otimes \mathbb{Z}_\ell \xrightarrow{\sim} Alb(X)(k)^\wedge$. We remark that $T(\bar{X})$ is uniquely ℓ-divisible by Roitman's theorem, in particular, $A_0(\bar{X})^{G_k} \longrightarrow Alb(X)(k)$ is surjective after tensoring with $\mathbb{Z}_\ell$.

This shows that the ℓ-adic Abel-Jacobi map can be non-injective, even up to torsion, for fields k of higher transcendence degree, while, on the other hand, the conjecture of Bloch and Beilinson (9.12) (implying $T(X) \otimes \mathbb{Q} = 0$ for a variety X/k) would imply the following conjecture for a number field k and $j = d$:

<u>9.15. Conjecture</u> Let X be smooth and proper over a finite or global field k, then for $\ell \neq char\ k$ the map

$$cl' \otimes \mathbb{Q}_\ell : CH^j(X)_0 \otimes \mathbb{Q}_\ell \longrightarrow H^1_{cont}(G_k, H^{2j-1}(\bar{X}, \mathbb{Q}_\ell(j)))$$

induced by the Abel-Jacobi map 9.4.1 is injective.

Before we discuss some examples, we investigate the relationship between conjecture 9.15 and the surjectivity of $ch_{2j-1,j}$ more closely. If in 9.1 we drop the assumption that Z is of codimension j, we get the following refined version.

We place ourselves again in the situation of a general weighted F-linear twisted Poincaré duality theory (F a field of characteristic zero), which receives Chern classes and satisfies condition p). Let X be a smooth, proper variety of pure dimension d, let $Z \subseteq X$ be a closed subscheme and let $U = X \setminus Z$. Let

$$N^Z H^i(X,j) = \mathrm{Im}(H^i_Z(X,j) \longrightarrow H^i(X,j)) ,$$

$$N^Z CH^r(X) = \mathrm{Im}(CH_{d-r}(Z)_0 \longrightarrow CH_{d-r}(X) = CH^r(X)) ,$$

where now $CH_m(Z)_0 = \mathrm{Ker}(CH_m(Z) \xrightarrow{cl} \Gamma H_{2m}(Z,m))$. Since for a cycle z of codimension j on X with support $Z_0 \subseteq Z$ we have a commutative diagram

$$\begin{array}{ccccccccc}
0 \longrightarrow & H^{2j-1}(X,j)/N^Z & \longrightarrow & H^{2j-1}(U,j) & \longrightarrow & H^{2j}_Z(X,j) & \longrightarrow & H^{2j}(X,j) \\
& \uparrow\!\!\uparrow & & \uparrow & & \uparrow & & \| \\
0 \longrightarrow & H^{2j-1}(X,j) & \longrightarrow & H^{2j-1}(X \setminus Z_0,j) & \longrightarrow & H^{2j}_{Z_0}(X,j) & \longrightarrow & H^{2j}(X,j) ,
\end{array}$$

we obtain a commutative exact diagram

$$\begin{array}{ccccccc}
0 \longrightarrow & \Gamma H^{2j-1}(U,j) & \longrightarrow & \Gamma H^{2j}_Z(X,j)_0 & \longrightarrow & R^1\Gamma(H^{2j-1}(X,j)/N^Z) \\
& \uparrow{\scriptstyle ch_{2j-1,j}} & & \uparrow{\scriptstyle cl} & & \uparrow{\scriptstyle cl'} \\
& K_1(U)^{(j)} & \longrightarrow & (CH^j_Z(X)_0/CH_{d-j}(Z)_0)_{\mathbb{Q}} & \longrightarrow & (CH^j(X)_0/N^Z)_{\mathbb{Q}} .
\end{array}$$

<u>9.16. Corollary</u> a) If $ch_{2j-1,j} \otimes F$ is surjective and

(9.16.1) $$cl \otimes F : (CH^j_Z(X)_0/CH_{d-j}(Z)_0) \otimes F \longrightarrow \Gamma H^{2j}_Z(X,j)_0$$

is still injective, then

(9.16.2) $$cl' \otimes F : (CH^j(X)_0/N^Z) \otimes F \longrightarrow R^1\Gamma(H^{2j-1}(X,j)/N^Z)$$

is injective on the subgroup generated by the cycles supported on Z .

b) Conversely, if 9.16.2 is injective on the subgroup supported on Z and 9.16.1 is surjective, then $ch_{2j-1,j} \otimes F$ is surjective.

For the following we observe that by definition

$$N^\nu H^i(X,j) = \bigcup_{Z \subset X \text{ of codimension } \nu} N^Z H^i(X,j)$$

is the coniveau filtration, while

$$N^\nu CH^j(X) = \bigcup_{Z \subset X \text{ of codimension } \nu} N^Z CH^j(X)$$

$= \{z \in CH^j(X) \mid \exists\ Z \subset X$ of codimension ν such that z is supported on Z and maps to zero in $\Gamma H_{2(d-j)}(Z,d-j)\}$

is the filtration described by Bloch and Ogus in [BO] (7.2). We can now state a refined version of conjecture 9.15.

<u>9.17. Conjecture</u> Let X be smooth and proper over a finite or global field k and let $\ell \neq \text{char } k$ be a prime.

a) Let $Z \subseteq X$ be a closed subscheme, then for all $j \geq 0$ the map

$$(9.17.1) \qquad (CH^j(X)_0/N^Z) \otimes \mathbb{Q}_\ell \longrightarrow H^1_{cont}(G_k, H^{2j-1}(\bar{X}, \mathbb{Q}_\ell(j))/N^Z)$$

induced by the Abel-Jacobi map is injective.

b) The Abel-Jacobi map induces injective maps

$$(9.17.2) \qquad Gr^\nu_N CH^j(X) \otimes \mathbb{Q}_\ell \longrightarrow H^1_{cont}(G_k, Gr^\nu_N H^{2j-1}(X, \mathbb{Q}_\ell(j)))$$

for all $j \geq 0$ and all $0 \leq \nu \leq j$.

<u>9.18. Lemma</u> a) Conjecture 9.17 a) implies conjecture 9.17 b).
b) Conjecture 9.17 b) implies conjecture 9.15.
c) If Tate's conjecture B (see 7.13) is true for cycles of dimension d - j on (possibly singular) closed subscheme of X, then conjecture 5.19 for $i = 2j - 1$ and open subvarieties of X is equivalent to conjecture 9.17 a) for j and X .
d) If $Z \subseteq X$ has a good proper cover $\tilde{Z} \to Z$ (e.g., if char $k = 0$), if $N^Z H^{2j-1}(\bar{X}, \mathbb{Q}_\ell)$ is a direct factor of $H^{2j-1}(\bar{X}, \mathbb{Q}_\ell)$ (e.g., if char $k = 0$), and if Tate's conjecture A is true for cycles of dimension $d = \dim X$ on $X \times \tilde{Z}$, then conjecture 9.15 for X implies conjecture 9.17 a) for X and Z .

<u>Proof</u> a) Conjecture 9.17 a) implies the injectivity of

$$(9.18.3) \qquad (CH^j(X)_0/N^\mu) \otimes \mathbb{Q}_\ell \longrightarrow H^1_{cont}(G_k, H^{2j-1}(\bar{X}, \mathbb{Q}_\ell(j))/N^\mu)$$

by passing to the limit over all $Z \subset X$ of codimension μ - note that for the ℓ-adic cohomology this limit is actually finite since $H^{2j-1} := H^{2j-1}(\bar{X}, \mathbb{Q}_\ell(j))$ is a finite-dimensional $\mathbb{Q}_\ell$-vector space. Since H^{2j-1} is pure of weight -1, $V^{G_k} = 0$ for every subquotient V of it; hence the top row in the commutative diagram

$$\begin{array}{ccccc} 0\to H^1_{cont}(G_k,Gr^{\nu}_N H^{2j-1}) & \to & H^1_{cont}(G_k,H^{2j-1}/N^{\nu+1}) & \longrightarrow & H^1_{cont}(G_k,H^{2j-1}/N^{\nu}) \\ \uparrow\alpha^{\nu} & & \uparrow & & \uparrow \\ 0\to Gr^{\nu}_N CH^j(X)_0\otimes\mathbb{Q}_\ell & \longrightarrow & (CH^j(X)_0/N^{\nu+1})\otimes\mathbb{Q}_\ell & \longrightarrow & (CH^j(X)_0/N^{\nu})\otimes\mathbb{Q}_\ell\to 0 \end{array} \tag{9.18.4}$$

is exact, and the map α^{ν} is well-defined (compare [Bl 3] 1.5), and injective if 9.18.3 is injective for $\nu + 1$.

b) Using 9.18.4, one shows the injectivity of 9.18.3 for all ν by induction on ν.

c) Tate's conjecture B for Z and cycles of dimension $d - j$ just assures the bijectivity of 9.16.1 for the ℓ-adic realization ($F = \mathbb{Q}_\ell$). Thus by 9.16 we only have to show that conjecture 5.19 for $(2j - 1,j)$ and all $U \subseteq X$ implies the injectivity of 9.17.1. Let $\alpha \in CH^j(X)_0$. If α is supported on Z, we are done by 9.16. Otherwise let α be supported on Z_0 which is of codimension j in X, and let $Z' = Z \cup Z_0$. Then obviously we have $N^Z CH^j(X) = N^{Z'} CH^j(X)$, and we may apply 9.16 to Z' and $U' = X \smallsetminus Z'$.

d) If $Z \xrightarrow{\iota} X$ has the good proper cover $\tilde{Z} \xrightarrow{\pi} Z$, we have
$N^Z H^{2j-1}(\bar{X},\mathbb{Q}_\ell(j))$
$= \mathrm{Im}(H_{2(d-j)+1}(\tilde{\bar{Z}},\mathbb{Q}_\ell(d-j)) \xrightarrow{\iota_*\pi_*} H_{2(d-j)+1}(\bar{X},\mathbb{Q}_\ell(d-j))$, and by the assumptions this is a direct factor of both the source and the target of $\iota_*\pi_*$ (for char k = 0 the latter follows via polarizations for absolute Hodge cycles as in §7). Thus there are sections s and t as indicated,

$$H_{2(d-j)+1}(\bar{\tilde{Z}},\mathbb{Q}_\ell(d-j)) \xrightarrow[\curvearrowleft t]{} N^Z \xrightarrow[\curvearrowleft s]{} H_{2(d-j)+1}(\bar{X},\mathbb{Q}_\ell(d-j)).$$

The composition ts can - via Poincaré duality and Künneth formula - be interpreted as an element in

$$H_{2d}(\overline{\tilde{Z} \times_k X}, \mathbb{Q}_\ell(d))^{G_k} .$$

By Tate's conjecture it would be the cycle class of a cycle W of dimension d on $\tilde{Z} \times_k X$. As a correspondence from X to $\tilde{Z}$ it just induces the map ts in the ℓ-adic cohomology, but it also gives us a map of the corresponding Chow groups. Composing with $\iota_* \pi_*$ we get the maps w and w' in the commutative exact diagram

$$\begin{array}{ccccccccc}
0 \to & H^1_{cont}(G_k, N^Z) & \overset{w}{\leftarrow\!\!\rightarrow} & H^1_{cont}(G_k, H^{2j-1}) & \to & H^1_{cont}(G_k, H^{2j-1}/N^Z) & \to 0 \\
& \uparrow & & \uparrow \scriptstyle{cl' \otimes \mathbb{Q}_\ell} & & \uparrow & \\
0 \to & N^Z CH^j(X) \otimes \mathbb{Q}_\ell & \overset{w'}{\leftarrow\!\!\rightarrow} & CH^j(X)_0 \otimes \mathbb{Q}_\ell & \to & (CH^j(X)_0 / N^Z CH^j(X)) \otimes \mathbb{Q}_\ell & \to 0
\end{array}$$

so that w is a section and w' is compatible with w. Although we don't know whether w' is a section, its compatibility with w suffices to deduce the injectivity of the right vertical map from the injectivity of $cl' \otimes \mathbb{Q}_\ell$.

Of course, a similar result holds for the Hodge realizations, and we get

<u>9.19. Corollary</u> Let X be a smooth and proper variety of dimension d over a number field. The following statements are equivalent.

i) Conjecture 5.19 (resp. 5.20) is true for $(i,j) = (2d - 1,d)$ and all open subvarieties U of X.

ii) The Abel-Jacobi map $cl' \otimes \mathbb{Q}_\ell$ (resp. $cl' \otimes \mathbb{Q}$) is injective for zero cycles on X.

iii) $T(X) \otimes \mathbb{Q} = 0$.

Moreover, conjecture 5.20 is true for $(i,j) = (3,2)$ and all open subvarieties U of X if and only if $cl' \otimes \mathbb{Q}$ is injective for cycles of codimension 2 on X.

Proof The Hodge conjecture and the Tate conjecture B) are true for zero cycles, hence i) is equivalent to 9.17 a) for $j = d$. We have to show that this is implied by ii). Since $CH^d(x)_O = N^{d-1}CH^d(X)_O$, one easily reduces to the case $\dim Z = 1$. Then we conclude by the absolute Hodge analogue of 9.18 d), since we can choose $\tilde{Z}$ to be a smooth curve, and the Hodge conjecture is true for divisors. For the equivalence between ii) and iii) observe that in both cases $T(X)$ is the kernel of the Abel-Jacobi map on $A_O(X)$. The last claim follows from the Hodge analogue of 9.18 c) and the fact that the Hodge conjecture is true for codimension 0 and 1 (trivial for 0 and Lefschetz' theorem plus (the proof of) 7.9).

9.20 Examples a) The case of a finite field k is discussed in chapter 12.

b) Let X be a smooth and proper variety over a finitely generated field k and let $A = \underline{Pic}^O(X)$ be the Picard variety of X. The exact Kummer sequences

$$0 \longrightarrow {}_{\ell^n}A \longrightarrow A \xrightarrow{\ell^n} A \longrightarrow 0$$

$(\ell \neq \text{char } k)$ induce the exact cohomology sequences

$$A(k) \xrightarrow{\ell^n} A(k) \xrightarrow{\delta} H^1(k, {}_{\ell^n}A) \longrightarrow {}_{\ell^n}H^1(k,A) \longrightarrow 0$$

for all $n \in \mathbb{N}$ and hence the exact sequence

$$0 \longrightarrow \varprojlim_n A(k)/\ell^n \xrightarrow{\delta} H^1_{cont}(G_k, T_\ell A) \longrightarrow T_\ell H^1(k,A) \longrightarrow 0 ,$$

where $T_\ell B = \varprojlim_n {}_{\ell^n}B$ is the Tate module of a group B. By the theorem of Mordell-Weil $A(k)$ is a finitely generated abelian group, hence $A(k) \otimes \mathbb{Z}_\ell \xrightarrow{\sim} \varprojlim_n A(k)/\ell^n$. Furthermore we have a canonical isomorphism $T_\ell A = T_\ell \mathrm{Pic}(\bar{X}) = H^1_{et}(\bar{X}, \mathbb{Z}_\ell(1))$ via the isomorphism of exact sequences

$$\begin{array}{ccccccccc} 0 \longrightarrow & H^1_{et}(\bar{X}, \mu_{\ell^n}) & \longrightarrow & H^1_{et}(\bar{X}, \mathbb{G}_m) & \xrightarrow{\ell^n} & H^1_{et}(\bar{X}, \mathbb{G}_m) & \longrightarrow 0 \\ & \wr\Vert & & \wr\Vert & & \wr\Vert & \\ 0 \longrightarrow & {}_{\ell^n}A(\bar{k}) & \longrightarrow & A(\bar{k}) & \xrightarrow{\ell^n} & A(\bar{k}) & \longrightarrow 0 , \end{array}$$

and via these identifications, δ agrees with the ℓ-adic Abel-Jacobi map

$$cl' \otimes \mathbb{Z}_\ell : CH^1(X)_0 \otimes \mathbb{Z}_\ell = \mathrm{Pic}^0(X) \otimes \mathbb{Z}_\ell \longrightarrow H^1_{cont}(G_k, H^1(\bar{X}, \mathbb{Z}_\ell(1))),$$

see [J 5]. Hence the latter is injective, and the commutative exact diagram (compare 9.8.1)

$$\begin{array}{ccccccc} 0 \longrightarrow H^1(\bar{U}, \mathbb{Z}_\ell(1))^{G_k} & \longrightarrow & H^2_{\bar{Z}}(\bar{X}, \mathbb{Z}_\ell(1))_0^{G_k} & \longrightarrow & H^1_{cont}(G_k, H^1(\bar{X}, \mathbb{Z}_\ell(1))) \\ \uparrow c_{1,1} & & \wr\uparrow & & \uparrow cl' \otimes \mathbb{Z}_\ell \\ 0 \longrightarrow (\mathcal{O}(U)^\times/k^\times) \otimes \mathbb{Z}_\ell & \longrightarrow & (\bigoplus_{x \in X^{(1)} \cap Z} \mathbb{Z}_\ell)_0 & \longrightarrow & CH^1(X)_0 \otimes \mathbb{Z}_\ell , \end{array}$$

for $Z \subseteq X$ closed and $U = X \smallsetminus Z$, gives another proof of 5.15 b), at least in the case that U has a smooth compactification.

c) Let X be a rational variety of dimension d (smooth, projective) over the global field k. Then $\mathrm{Pic}(\bar{X})$ is a torsion-free, finitely generated abelian group and hence

$H^1(\bar{X},\mathbb{Q}_\ell/\mathbb{Z}_\ell(1)) = 0$. By duality $H^{2d-1}(\bar{X},\mathbb{Z}_\ell(d)) = 0$, so conjectures 9.15 and 9.17 are true for $j = d$ if and only if $A_0(X) = CH^d(X)_0$ is a torsion group. For $d = 2$ Colliot-Thélène [C] in fact proved that $A_0(X)$ is finite, except possibly for p-torsion for the case that $\mathrm{char}\, k = p > 0$.

d) in [Bl 3] Bloch considers $CH^2(X)$ for a certain threefold X over a number field to test the generalized conjecture of Birch and Swinnerton-Dyer (see 9.20.1 below). He exhibits cases where the L-function of $Gr_N^0 H^3(\bar{X},\mathbb{Q}_\ell)$ has a zero at $s = 3$ and at the same time there is a non-zero cycle Ξ in $Gr_N^0 CH^2(X)_{\mathbb{Q}}$, which is the group of cycles homologous to zero modulo those which are algebraically equivalent to zero (tensored with $\mathbb{Q}$), see loc. cit. Lemma 1.1 and below. To prove that Ξ is non-zero Bloch in fact proves that its image in $H^1_{cont}(G_k, Gr_N^0 H^3(\bar{X},\mathbb{Q}_\ell(2)))$ under the Abel-Jacobi map is non-zero.

e) Bloch [Bl 4] and Beilinson [Bei 4] independently proposed the following conjecture for a smooth projective variety X over a number field k, generalizing the conjecture of Birch and Swinnerton-Dyer :

<u>9.20.1. Conjecture</u> $\mathrm{rk}\, Gr_N^\nu CH^j(X) = \mathrm{ord}_{s=j} L(Gr_N^\nu H^{2j-1}(\bar{X},\mathbb{Q}_\ell),s)$.

Bloch proves that $Gr_N^{j-1} CH^j(X) = N^{j-1} CH^j(X)$ is the group of cycles algebraically equivalent to zero and that for this part the conjecture is strongly related to the original conjecture of Birch and Swinnerton-Dyer for abelian varieties (which is the case $j = 1$). With the same arguments we can prove

9.20.2. Lemma Assume that the Birch and Swinnerton-Dyer conjecture

$$(9.20.3) \qquad \operatorname{rk} A(k) = \operatorname{ord}_{s=1} L(T_\ell(A) \otimes_{\mathbb{Z}_\ell} \mathbb{Q}_\ell, s)$$

for abelian varieties A over the number field k is true. Then conjecture 9.20.1 for $\nu = j - 1$ is equivalent to conjecture 9.17 b) for $\nu = j - 1$, i.e., to the injectivity of the ℓ-adic Abel-Jacobi map up to torsion on the subgroup of cycles algebraically equivalent to zero.

Proof In [Bl 4] 1.4 Bloch constructs an abelian variety W with rational Tate module $T_\ell W \otimes \mathbb{Q}_\ell = Gr_N^{j-1} H^{2j-1}(\bar{X}, \mathbb{Q}_\ell(j))$ together with a surjection $Gr_N^{j-1} CH^j(X)_{\mathbb{Q}} \twoheadrightarrow W(k)_{\mathbb{Q}}$. It follows from the construction that the diagram

$$\begin{array}{ccc} W(k) \otimes \mathbb{Q}_\ell & \overset{\delta}{\hookrightarrow} & H^1_{cont}(G_k, T_\ell W \otimes \mathbb{Q}_\ell) \\ \uparrow & & \wr\| \\ Gr_N^{j-1} CH^1(X) \otimes \mathbb{Q}_\ell & \xrightarrow{cl'} & H^1_{cont}(G_k, Gr_N^{j-1} H^{2j-1}(X, \mathbb{Q}_\ell(j))) \end{array}$$

is commutative, where cl' comes from the Abel-Jacobi map and δ is induced by Kummer sequence for B and hence is injective (see b) above; for the commutativity see [J5]). As Bloch points out, under assumption of 9.20.3 for W, conjecture 9.20.1 for $\nu = j - 1$ is equivalent to the bijectivity of the left vertical map, so the claim follows.

§10. On the non-injectivity of the Abel-Jacobi map

Let us first extend Bloch's theorem (9.13) to higher dimensions.

<u>10.1. Theorem</u> Let X be a connected, smooth, proper variety of dimension $d \geq 2$ over the field k, and let K be the function field of X. If $H^d(\bar{X},\mathbb{Q}_\ell) \neq N^1H^d(\bar{X},\mathbb{Q}_\ell)$ for the filtration by coniveau N^ν ($\ell \neq \mathrm{char}(k)$), then $T(X\times_k K) \otimes \mathbb{Q} \neq 0$.

<u>Proof</u> We closely follow the arguments in [Bl1] lecture 1, cf. also [BS]. First we show that, under the assumption made, the class of the diagonal $\Delta \subseteq X\times_k X$ does not restrict to zero under

$$CH^d(X\times_k X) \to CH^d(X\times_k K) \to CH^d(U'\times_k K)\otimes\mathbb{Q}$$

for any non-empty open $U' \subseteq X$.

In fact, this would imply that Δ restricts to zero in $CH^d(U'\times_k U)\otimes\mathbb{Q}$ for some non-empty open $U \subseteq X$, hence

$$N\cdot[\Delta] = [\Gamma_1] + [\Gamma_2]$$

for some $N \in \mathbb{N}$, a cycle Γ_1 supported on $D'\times X$ and a cycle Γ_2 supported on $X\times D$, for some divisors D, D' on X. It is easy to see that the correspondence induced by Γ_2 maps $H^d(\bar{X},\mathbb{Q}_\ell)$ to $\mathrm{Im}(H_d(\bar{D},\mathbb{Q}_\ell(d)) \to H_d(\bar{X},\mathbb{Q}_\ell(d))) \subseteq N^1H^d(\bar{X},\mathbb{Q}_\ell)$, see loc. cit. . Now let $Z \subset D'\times X$ be irreducible and in the support of Γ_1. Let D'' be the image of Z under $\pi_1: Z \to D'\times X \xrightarrow{pr_1} D'$, let $V_1 \subset D''$ be open, non-empty, smooth and affine, and let $Y_1 = D''\setminus V_1$. Then $H^d(\bar{V}_1,\mathbb{Q}_\ell) = 0$, since $\dim V_1 < d$ ([Mi] 7.2). If $\pi_2: Z \to D'\times X \xrightarrow{pr_2} X$ is not surjective, we can conclude as before that the correspondence given by Z is zero on $H^d(\bar{X},\mathbb{Q}_\ell)/N^1H^d(\bar{X},\mathbb{Q}_\ell)$. If π_2 is surjective, let $Y_2 = \pi_1^{-1}(Y_1)$, $Y_3 = \pi_2(Y_2)$, $V_3 = X\setminus Y_3$ and $V_2 = \pi_2^{-1}(V_3)$.

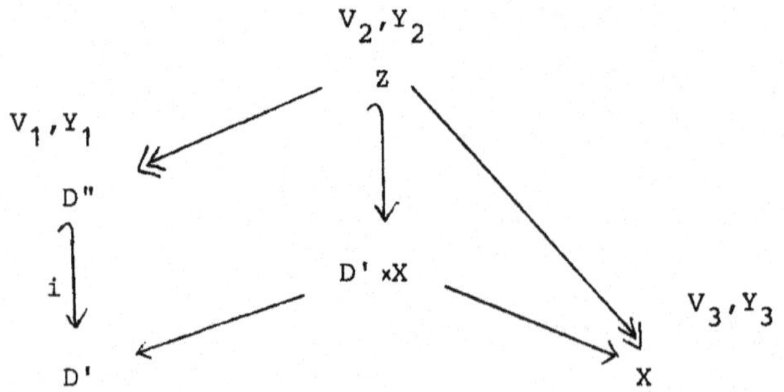

Since $\dim Z = d$, we have $\dim Y_3 < d$, hence $V_3 \neq \emptyset$. By the projection formula, the correspondence given by Z is the bottom way in the commutative diagram

$$\begin{array}{ccccccc}
 & \nearrow & H^d(\bar V_2,\mathbb{Q}_\ell) & \xrightarrow{\cap\eta_{\bar V_2}} & H_d(\bar V_2,\mathbb{Q}_\ell(d)) & \searrow & \\
0 = H^d(\bar V_1,\mathbb{Q}_\ell) & & \uparrow & & \uparrow & & H_d(\bar V_3,\mathbb{Q}_\ell) \\
\uparrow & & H^d(\bar Z,\mathbb{Q}_\ell) & \xrightarrow{\cap\eta_{\bar Z}} & H_d(\bar Z,\mathbb{Q}_\ell(d)) & (\pi_1)_* & \uparrow \\
H^d(\bar D'',\mathbb{Q}_\ell) & \overset{\pi_1^*}{\nearrow} & & & & \searrow & H_d(\bar X,\mathbb{Q}_\ell(d)) \\
i^*\uparrow & & & & & & \wr \\
H^d(\bar X,\mathbb{Q}_\ell) & & & & & & H^d(\bar X,\mathbb{Q}_\ell)
\end{array}$$

hence its composition with $H^d(\bar X,\mathbb{Q}_\ell) \to H^d(\bar V_3,\mathbb{Q}_\ell)$ is zero. We find that Γ_1 maps $H^d(\bar X,\mathbb{Q}_\ell)$ to $N^1H^d(\bar X,\mathbb{Q}_\ell)$, too. Since Δ acts as the identity, we would conclude $H^d(\bar X,\mathbb{Q}_\ell) = N^1H^d(\bar X,\mathbb{Q}_\ell)$, contrary to the assumption.

To prove the theorem choose a smooth proper curve and a morphism $f\colon C \to X$ such that the composition

$$\operatorname{Pic}^o(C)(K)\otimes\mathbb{Q} \xrightarrow{f_*} A_o(X\times_k K)\otimes\mathbb{Q} \to A\ell b(X)(K)\otimes\mathbb{Q}$$

is surjective, and let $\eta \in CH_o(X\times_k K)$ be the 0-cycle corresponding to the generic point $\operatorname{Spec} K \to X$. Then there is a $c \in \operatorname{Pic}(C)$ such that $\eta - f_*(c)$ is mapped to zero in $A\ell b(X)(k)\otimes\mathbb{Q}$, and for any such c the element $\eta - f_*(c) \in T(X\times_k K)\otimes\mathbb{Q}$ is non-zero, since $f_*(c)$ restricts to zero in $CH_o(U'\times_k K)\otimes\mathbb{Q}$ for $U' = X\setminus f(c)$, but η does not, as the image of Δ.

<u>10.2. Remarks</u> a) If k is uncountable and algebraically closed, then one may replace "$T(X\times_k K)\otimes\mathbb{Q} \neq 0$" by "$T(X)\otimes\mathbb{Q} \neq 0$", by applying everything to a model X_o of X over a finitely generated field k_o and using the inclusion $T(X_o\times_{k_o} K_o)\otimes\mathbb{Q} \hookrightarrow T(X)\otimes\mathbb{Q}$ induced by an embedding $K_o \hookrightarrow k$ of the function field of X_o into k (cf. [Bl 1]).

b) For a field k of characteristic zero one has

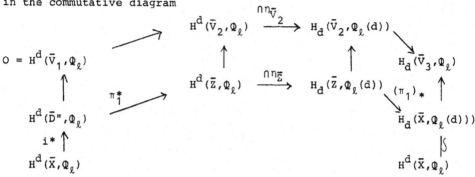

(10.2.1) $H^o(X,\Omega^i_X) \neq 0 \Rightarrow N^1H^i(\bar{X},\mathbb{Q}_\ell) \neq H^i(\bar{X},\mathbb{Q}_\ell)$. In fact, by a similar argument as in a) and specialization of cycles it suffices to consider the case that k is embeddable in $\mathbb{C}$, and then $k = \mathbb{C}$, where we may replace $H^i(X,\mathbb{Q}_\ell)$ by $H^i(X(\mathbb{C}),\mathbb{Q})$. Then the claim follows from Hodge theory, as a special case of the investigations in [Gr]. There Grothendieck defines a filtration $'N^\nu$ by setting

$$'N^\nu H^i(X,\mathbb{Q}) = \begin{array}{l}\text{union of the sub-Hodge structures } W, \text{ for} \\ \text{which } W(\nu) \text{ if effective ,}\end{array}$$

where effective Hodge structures are those with $p,q \geq 0$ for the occuring Hodge types (p,q). Grothendieck then shows the inclusion $N^\nu \subseteq {}'N^\nu$ - which obviously proves 10.2.1 - and states as the generalized Hodge conjecture that $N^\nu = {}'N^\nu$ - which would imply the converse of 10.2.1.

c) In [D.E.] VI 10.3 Grothendieck has formulated a generalized Tate conjecture for varieties over finite fields. This can be extended to finitely generated fields . Call a constructible , pure $\mathbb{Q}_\ell$-sheaf F of weight w on a scheme S of finite type over $\mathbb{Z}[\frac{1}{\ell}]$ <u>effective</u> if F is entire ([D9] 3.3.2) and if $F^\vee(w)$ is entire. For $\ell \neq \text{char } k$ call an object V in $\text{Rep}_c(G_k,\mathbb{Q}_\ell)$ entire or effective, if it extends to such a $\mathbb{Q}_\ell$-sheaf over a model U of k as in 6.8. For X/k smooth, projective $H^i(\bar{X},\mathbb{Q}_\ell)$ is effective (see [J3], proof of thm. 1), and we let

$$'N^\nu H^i(\bar{X},\mathbb{Q}_\ell) = \begin{array}{l}\text{union of the subrepresentation } W \text{ for} \\ \text{which } W(\nu) \text{ if effective} \quad .\end{array}$$

It follows from [D6] 3.3.8 that for a variety Y of dimension e over k the weight-graded parts of $H_a(\bar{Y},\mathbb{Q}_\ell(e)) = H^a_c(\bar{Y},\mathbb{Q}_\ell)^\vee(-e)$ are effective (note that a pure $\mathbb{Q}_\ell$-representation of $\text{Gal}(\bar{\mathbb{F}}_q/\mathbb{F}_q)$ is effective if and only if $r,s \geq 0$ for the occuring p-adic Hodge types (r,s) as defined in loc. cit. 3.3.7). Hence for $Y \subseteq X$ the G_k-representation

$$\text{Im}(H_a(\bar{Y},\mathbb{Q}_\ell(d)) \to H_a(\bar{X},\mathbb{Q}_\ell(d)) \cong H^{2d-a}(\bar{X},\mathbb{Q}_\ell))(d-e)$$

is effective, for d = dim X, which shows the inclusion $N^{\nu} \subseteq {}'N^{\nu}$. The generalized Tate conjecture states the equality $N^{\nu} = {}'N^{\nu}$.

<u>10.3</u> By Roitman's theorem (9.11.3) and theorem 10.1 one gets examples of non-injective complex or ℓ-adic Abel-Jacobi maps for 0-cycles on varieties of arbitrary dimension d. Several questions naturally arise.

The examples provided so far always used cycles defined over fields of higher transcendence degree than the field of definition for X, and one may ask whether this is essential. However, Schoen [Schoe] has produced a counterexample, where both the variety and the cycles are defined over a finite extension of $\mathbb{Q}(t)$ (see appendix B).

As far as I know there do not exist examples for cycles of higher dimension, but I think it should be possible to construct these, too, for example by the method described in 10.4 below.

Finally, one may wonder whether the Abel-Jacobi map could be injective for other theories, e.g., for the theory of absolute Hodge cycles. However, it turns out that the Abel-Jacobi map is non-injective quite principally in codimension $j \geq 2$ (over fields of higher transcendence degree), in view of the following construction.

<u>10.4. Principle</u> Let X be a smooth projective variety. If $z_1 \in CH^i(X)_o$ and $z_2 \in CH^j(X)_o$ are cycles which are homologous to zero, then their intersection $z_1 \cdot z_2 \in CH^{i+j}(X)_o$ lies in the kernel of the Abel-Jacobi map.

We can turn this principle into a theorem in the following two cases, the first one explaining the idea behind 10.4 and probably applying for all theories of interest.

10.5. Assume that the cohomology theory $H^*(X,j)$ is obtained as the homology of some complex $\underline{R\Gamma}(X,j)$ in $D^b(T)$ for our tensor category T (not necessarily equipped with weights), and that the cycle map factorizes through some "absolute cohomology" - the homology of

(10.5.1) $R\Gamma_T\underline{R\Gamma}(X,j) = R\,\mathrm{Hom}_{D^b(T)}(1,\underline{R\Gamma}(X,j))$;

this is fulfilled for the ℓ-adic theory ($T = S((\mathrm{Spec}\ k)_{\acute{e}t})^{\mathbb{Z}_\ell}$ cf. 9.6) and the Hodge theory ($T = MH$, cf. 9.7 c)) . Then by 9.5 the Abel-Jacobi map is induced by the hypercohomology spectral sequence

(10.5.2) $E_2^{p,q} = R^p\Gamma_T H^q(X,j) \Rightarrow H_T^{p+q}(X,j)$,

where by definition $H_T^i(X,j) = \mathrm{Ext}^i(1,\underline{R\Gamma}(X,j))$ = i-th homology of 10.5.1. If there is a cupproduct compatible with this spectral sequence, e.g., if it comes from a pairing

(10.5.3) $\underline{R\Gamma}(X,i) \otimes^L \underline{R\Gamma}(X,j) \to \underline{R\Gamma}(X,i+j)$,

then one has

(10.5.4) $F^r H_T^m(X,i) \otimes F^s H_T^n(X,j) \to F^{r+s} H_T^{m+n}(X,i+j)$

for the descending filtration F^ν on the limit terms associated to 10.5.2. If now z_1, z_2 are homologically equivalent for $H^*(X,*)$, then by definition $cl(z_1) \in F^1H_T^{2i}(X,i)$ and $cl(z_2) \in F^1H_T^{2j}(X,j)$ for the absolute cycle classes. If the cupproduct is compatible with the intersection product, then 10.5.4 implies $cl(z_1\cdot z_2) \in F^2H_T^{2(i+j)}(X,i+j)$, hence it is mapped to zero in $R^1\Gamma_T H^{2(i+j)-1}(X,i+j)$ by **definition** of F^ν . For the Hodge and the ℓ-adic theory such a cupproduct exists (see [Bei 2] 4.2 and [J1] §6), and hence principle 10.4 is a theorem for these. For absolute Hodge cycles one could proceed in a similar way, by using the absolute Hodge complexes of 6.11; instead we shall deduce it by a different method, also valid for more general cohomology theories.

10.6. To prove 10.4 without using an absolute cohomology theory I need the following compatibility for the considered twisted Poincaré duality theory.

r) For $Y \subset X$ closed and $U = X \smallsetminus Y$ the following diagram of natural maps commutes

$$\begin{array}{ccc} H_a(X,b) \otimes H^i_Y(X,j) & \xrightarrow{\cap} & H_{a-i}(Y,b-j) \\ \downarrow \qquad \uparrow\delta & & \uparrow\delta \\ H_a(U,b) \otimes H^{i-1}(U,j) & \xrightarrow{\cap} & H_{a-i+1}(U,b-j) \end{array} .$$

s) For X smooth and projective of pure dimension d the cycle map is compatible with the capproduct in the following sense: For cycles Z_1 and Z_2 of dimension i and j on X intersecting properly the diagram

$$\begin{array}{ccccc} \Gamma H_{2(d-i)}(Z_1,d-i) & \otimes & \Gamma H^{2j}_{Z_2}(X,j) & \xrightarrow{\cap} & \Gamma H_{2(d-i-j)}(Z_1 \cap Z_2,d-i-j) \\ \uparrow cl & & \uparrow cl & & \uparrow cl \\ Z^0(Z_1) & \otimes & Z^0(Z_2) & \xrightarrow{\text{intersection}} & Z^0(Z_1 \cap Z_2) \end{array}$$

commutes (Here the upper map is induced by the restriction $H^{2j}_{Z_2}(X,j) \rightarrow H^{2j}_{Z_1 \cap Z_2}(Z_1,j)$ and the capproduct).

For example, both things hold true for the absolute Hodge theory 6.11: this can be checked in any of the realizations, and is easily checked for the ℓ-adic theory.

Let z_1, z_2 be as in 10.4, and let Z_ν be the support of z_ν, $\nu = 1,2$. Assuming r) and s), and that $\otimes$ is an exact functor (this is true, e.g., if $\mathcal{T}$ is a rigid tensor category [DMOS] II 1.16), we get a commutative diagram with exact rows

$$\begin{array}{ccccccc}
0 \to & H_{2(d-i)}(Z_1)\otimes H^{2j-1}(X) & \to & H_{2(d-i)}(Z_1)\otimes H^{2j-1}(X\setminus Z_2) & \overset{id\otimes\delta}{\to} & H_{2(d-i)}(Z_1)\otimes H^{2j}_{Z_2}(X) \\
& \downarrow & & \downarrow & & \downarrow \\
0 \to & H_{2(d-i)}(Z_1)\otimes H^{2j-1}(Z_1) & \to & H_{2(d-i)}(Z_1)\otimes H^{2j-1}(Z_1\setminus Z) & \overset{id\otimes\delta}{\to} & H_{2(d-i)}(Z_1)\otimes H^{2j}_{Z}(Z) \\
& & & \downarrow & & \\
& \Big\downarrow \cap & & H_{2(d-i)}(Z_1\setminus Z)\otimes H^{2j-1}(Z_1\setminus Z) & & \Big\downarrow \cap \\
& & & \downarrow \cap & & \\
0 \to & H_{2(d-i-j)+1}(Z_1) & \longrightarrow & H_{2(d-i-j)+1}(Z_1\setminus Z) & \longrightarrow & H_{2(d-i-j)}(Z) \\
& \downarrow & & \downarrow & & \| \\
0 \to & H_{2(d-i-j)+1}(X) & \longrightarrow & H_{2(d-i-j)+1}(X\setminus Z) & \longrightarrow & H_{2(d-i-j)}(Z)\ ,
\end{array}$$

where we have omitted the twists and set $Z = Z_1 \cap Z_2$. Hence we get a commutative diagram with exact rows

$$\begin{array}{ccccccc}
0\to & H_{2(d-i)}(Z_1)_o\otimes H^{2j-1}(X) & \to & H_{2(d-i)}(Z_1)_o\otimes H^{2j-1}(X\setminus Z_2) & \to & H_{2(d-i)}(Z_1)_o\otimes H^{2j}_{Z_2}(X)_o & \to 0 \\
(10.6.1) & \Big\downarrow 0 & & \Big\downarrow & & \Big\downarrow \cap & \\
0\to & H_{2(d-i-j)+1}(X) & \to & H_{2(d-i-j)+1}(X\setminus Z) & \to & H_{2(d-i-j)}(Z)_o & \to 0
\end{array}$$

where $(\ )_o$ means "homologous to zero on X" and the vanishing of the left maps follows via the projection formula. Thus, the pull-back of the bottom exact sequence via $\cap$ is trivial, and, consequently, the same is true for the pull-back via $cl(z_1\cdot z_2)$ $\in \mathrm{Hom}_T(1,H_{2(d-i-j)}(Z,d-i-j)_o)$, since by s) it factorizes through $\cap$. But the pull-back by $cl(z_1\cdot z_2)$ by definition is the image of $z_1\cdot z_2$ in $\mathrm{Ext}^1_T(1,H^{2i+2j-1}(X,i+j))$ under the Abel-Jacobi map, which proves 10.4.

<u>10.7.</u> To deduce the non-injectivity of the Abel-Jacobi map from 10.4 it remains to find elements z_1, z_2 as there such that $z_1\cdot z_2$ is non-zero in $CH^{i+j}(X)\otimes\mathbb{Q}$. First note that cycles in $\mathrm{Pic}^o(X)$ are homologous to zero, if this is true for all smooth, projective

curves $C \to X$ (use that elements in $\mathrm{Pic}^o(X)$ come from Pic^o of curves via algebraic correspondences). This will be the case for all "geometric" theories; in any case it is true under each of the following conditions: a) $H^2(U,1) = 0$ for affine curves, b) $H^*(-,*)$ is $\mathbb{Q}$-rational, and $H^2(\mathbb{A}^1_k,1) = 0 = H^1(\mathrm{Spec}\ k,0)$. Of course, it is true for the theory of absolut Hodge cycles.

Assuming $\mathrm{Pic}^o(X) \subseteq CH^1(X)_o$, we may use Bloch's result that for an abelian surface X over an algebraically closed field k the map

(10.7.1) $\quad \mathrm{Pic}^o(X) \times \mathrm{Pic}^o(X) \to T(X)$

induced by intersection is surjective, see [Bl 4] . For char(k) = 0 one has $T(X\times_k K)\otimes\mathbb{Q} \neq 0$ for suitable field extensions by Roitman's theorem (9.11.3), since $H^2(X,\mathcal{O}_X) = \Lambda^2 H^1(X,\mathcal{O}_X) \neq 0$. As an example in positive characteristic one has $T(X\times_k K)\otimes\mathbb{Q} = 0$ by Bloch's theorem (9.13) for $X = E_1\times E_2$, E_i elliptic curves over $\mathbb{F}_p$, not both of them supersingular, and K the function field of X, since $H^2(\bar{X},\mathbb{Q}_\ell(1)) \neq H^2(\bar{X},\mathbb{Q}_\ell(1))^{G_k} = \tilde{N}^1 H^2(\bar{X},\mathbb{Q}_\ell(1))$, $\ell \neq p$.

<u>10.8. Remarks</u> a) Other examples where 10.7.1 is surjective and $p_g(X) > 0$ are Fano surfaces, see [Bl 1] 1.7.

b) In [Bl 1] lecture 1 Bloch conjectures that the converse of Mumford's or his theorem (see 9.11.2 and 9.13) is true, or, equivalently, that H^2 of a surface X is generated by algebraic cycles if and only if $A_o(X)$ is representable (by Alb(X); note that T(X) is torsion-free by Roitman's theorem [Ro 2]).

c) It is quite likely that T(X) lies in the kernel of every reasonable Abel-Jacobi map, but I have no proof for this.

d) Using Zarchin's proof of the Tate conjecture on endomorphisms of abelian varieties in characteristic $p > 0$, it is easy to show the following: If k is a finitely generated field of characteristic $p > 0$ and X is an abelian surface over k , then $H^2(\bar{X},\mathbb{Q}_\ell(1)) =$

$H^2(\bar{X},\mathbb{Q}_\ell(1))^{G_k}$ if and only if X is isogenous to a product of supersingular elliptic curves defined over $\mathbb{F}_p$. Hence only in this case one expects representability of $A_o(X)$ and an injective Abel-Jacobi map.

§11. Chow groups over arbitrary fields

In this section we discuss the structure of Chow groups expected over fields with arbitrary transcendence degree. For a smooth, projective surface X over $\mathbb{C}$, Bloch ([Bl 1] Lecture 1) considers a descending filtration F^ν on $CH^2(X)$ such that

$$Gr_F^o\ CH^2(X) = CH^2(X)/A_o(X) \overset{cl}{\underset{\sim}{\to}} H^4(X,\mathbb{Z}(2)) ,$$
$$Gr_F^1\ CH^2(X) = A_o(X)/T(X) \overset{cl'}{\underset{\sim}{\to}} H^3(X,\mathbb{C})/H^3(X,\mathbb{Z}(2)) + F^2 ,$$
$$Gr_F^2\ CH^2(X) = F^2CH^2(X) = T(X) ,$$

and conjectures a relation between $Gr_F^2CH^2(X)$ and H^2, cf. also his conjecture recalled in 10.8 b) above. He shows that the latter one would be true, if the action of algebraic correspondences on $Gr_F^{\cdot}CH^2(X)$ factorized through homological equivalence.

The following conjecture of Beilinson is a generalization of this.

<u>11.1. Conjecture</u> ([Bei 4] 5.10) Let k be a field. For smooth, projective varieties X over k there is a descending filtration F^ν on $CH^j(X)$

$$\ldots \subseteq F^{\nu+1}CH^j(X) \subseteq F^\nu CH^j(X) \subseteq \ldots \subseteq F^oCH^j(X) = CH^j(X) ,$$

such that

a) $F^\nu CH^j(X) = 0$ for $\nu >> 0$,

b) $F^1CH^j(X) = CH^j(X)_{num} = \{z \in CH^j(X) \mid z$ numerically equivalent to zero$\}$,

c) $F^rCH^i(X).F^sCH^j(X) \subseteq F^{r+s}CH^{i+j}(X)$ under the intersection product,

d) F^ν is functorial, i.e., respected by f^* and f_* for morphisms $f: X \to Y$,

e) (note that by c) and d) algebraic correspondences act on $Gr_F^{\cdot}CH^j(X)$ and that by b) this action factorizes through numerical equivalence) $Gr_F^{\nu}CH^j(X)\otimes\mathbb{Q}$ only depends on the Grothendieck motive w.r.t. numerical equivalence $H^{2j-\nu}(X)$.

The last statement is conditional, since the existence of the motive $H^{2j-\nu}(X)$, as a direct factor of the motive $H(X)$, see [Kl], depends on Grothendieck's standard conjectures that numerical equivalence equals homological equivalence (for ℓ-adic cohomology $H^*(\bar{X},\mathbb{Q}_\ell(*))$, $\ell \neq \text{char } k$, say) and that the Künneth components $\Delta(r,s)$ of the cycle class of the diagonal $\Delta \subset X\times X$ are algebraic. Part e) then means that

$$\Delta(r,s)\Big|_{Gr_F^{\nu}CH^j(X)_{\mathbb{Q}}} = \begin{cases} \text{id} , & (r,s)=(2d-2j+\nu,2j-\nu) , \\ 0 , & \text{otherwise} , \end{cases}$$

where d is the dimension of X (X connected, without restriction). The conditions a)-e) are not independent; for example, property e) automatically implies $Gr_F^{\nu} CH^j(X)_{\mathbb{Q}} = 0$ for $\nu > j$:

<u>11.2. Lemma</u> (Assuming the standard conjectures) Under assumptions b)-d), $\Delta(r,s)$ is zero on $Gr_F^{\cdot} CH^j(X)_{\mathbb{Q}}$ for $s < j$.

<u>Proof</u> (compare [Bl 1] 1.9) By the hard Lefschetz isomorphism $H^s(X) \cong H^{2d-s}(X)(d-s)$, the correspondence $\Delta(r,s)_*$ factorizes as $\Delta(r,s)_* = \Gamma'_* \circ \Gamma_*$, with $\Gamma \in CH^{2d-s}(X\times X)$ and $\Gamma' \in CH^s(X\times X)$; but then Γ_* has image in $CH^{d-s+j}(X)$:

$$\Gamma_* : \quad CH^j(X) \xrightarrow{pr_1^*} CH^j(X\times X) \xrightarrow{\cdot\Gamma} CH^{2d-s+j}(X\times X) \xrightarrow{(pr_2)_*} CH^{d-s+j}(X)$$

and this group vanishes for $s < j$.

11.3. There are in fact more precise conjectures on the origin of the above filtration F^ν. Namely, Beilinson conjectures that there should exist a suitable tensor category with weights $\mathcal{MM}$ and complexes $\underline{R\Gamma}(X,j) \in Ob(D^b(\mathcal{MM}))$ as in 10.5, which together with versions "with support" $\underline{R\Gamma_Z}(X,j)$ and homological counterparts $\underline{R\Gamma'}(X,b)$ satisfy certain axioms (in short, analogues in the derived category of the axioms of a twisted Poincaré duality theory, cf. the axioms in [Gi] and [Bei 1] 2.3, in particular pairings $\underline{R\Gamma}(X,i) \otimes^L \underline{R\Gamma}(X,j) \to \underline{R\Gamma}(X,i+j)$ inducing a cupproduct on the cohomology) so that

$$(X,Z) \rightsquigarrow H^i_{\mathcal{MM},Z}(X,j) := H^i(\underline{R\Gamma_Z}(X,j)) \in Ob(\mathcal{MM}) ,$$

$$X \rightsquigarrow H^{\mathcal{MM}}_a(X,b) := H_{-a}(\underline{R\Gamma'}(X,b)) \in Ob(\mathcal{MM})$$

forms a twisted Poincaré duality theory with weights, and such that for X smooth and projective the following holds:

a) The association $X \rightsquigarrow \bigoplus_{i\geq 0} H^i_{\mathcal{MM}}(X,0)\otimes\mathbb{Q} \in Ob(\mathcal{MM}\otimes\mathbb{Q})$ induces a weight-preserving equivalence of tensor categories between Grothendieck's category of $\mathbb{Q}$-motives over k (w.r.t. numerical equivalence) and the subcategory of semi-simple objects in $\mathcal{MM}$,

b) the cycle map induces an isomorphism

$$CH^j(X) \xrightarrow{\sim} H^{2j}_{\mathcal{MM}}(X,j) = Hom_{D^b(\mathcal{MM})}(1,\underline{R\Gamma}(X,j)[2j]),$$

c) and the (hyperext) spectral sequence (cf. 10.5.2)

$$E_2^{p,q} = R^p\Gamma_{\mathcal{MM}}H^q_{\mathcal{MM}}(X,j) \Rightarrow H^{p+q}_{\mathcal{MM}}(X,j)$$

degenerates after tensoring with $\mathbb{Q}$.

Then the filtration F^ν on $CH^j(X)$ would be defined by the spectral sequence c), which assures the properties 11.1 c) and d) provided the cupproduct is compatible with the intersection of cycles (compare 10.5). By a) we have $F^1CH^j(X)=Ker(CH^j(X)\to\Gamma H^{2j}_{\mathcal{MM}}(X,j))= CH^j(X)_{num}$, hence 11.1 b), and 11.1 e) is clear; in fact, by c) we have the formula

$$Gr^\nu_F CH^j(X)\otimes\mathbb{Q} \cong R^\nu\Gamma_{\mathcal{MM}}H^{2j-\nu}_{\mathcal{MM}}(X,j) \otimes \mathbb{Q} , \tag{11.3.1}$$

which makes more precise how $Gr_F^\nu CH^j(X)\otimes\mathbb{Q}$ depends on the motive $H^{2j-\nu}(X)$ $=H_{MM}^{2j-\nu}(X,0)$ (part of the axioms is the existence of a Tate object $1(1)$ in MM, cf. [DMOS] II §5, such that $\underline{R\Gamma}_Z(X,j) = \underline{R\Gamma}_Z(X,0)(j)$ and $\underline{R\Gamma'}(X,b) = \underline{R\Gamma'}(X,0)(-b)$; in particular $H_{MM}^{2j-\nu}(X,j) = H_{MM}^{2j-\nu}(X,0)(j))$.

A category MM as above could well be regarded as a category of mixed motives over k, and in generalization of b) above Beilinson expects a formula

(11.3.2) $\quad H_M^i(X,\mathbb{Q}(j)) \cong H_{MM}^i(X,j) \otimes \mathbb{Q}\ , \ i,j \in \mathbb{Z}$

for the motivic cohomology defined by K-theory, thus justifying its name. To deduce all this from Beilinson's formulation in [Bei 4], let $MM = M(\mathrm{Spec}\ k,\mathbb{Z})$ and $\underline{R\Gamma}(X,j) = Rf_*\mathbb{Z}_M(j)$ for $f:X\to\mathrm{Spec}\ k$ with Beilinson's notation in loc. cit., so that

$$\begin{aligned} H^i(X,\mathbb{Z}_M(j)) &= \mathrm{Ext}^i_{M(X,\mathbb{Z})}(\mathbb{Z}_M(0),\mathbb{Z}_M(j)) \\ &= \mathrm{Ext}^i_{MM}(1,Rf_*\mathbb{Z}_M(j)) = H^i_{MM}(X,j) \ . \end{aligned}$$

<u>11.4. Remarks</u> a) A similar interpretation of $K_m(X)^{(j)}$ in terms of a category of mixed motives over k was also given by Deligne [D10] .

b) The following axiom would work equally well as 11.3 a):

a') $X \rightsquigarrow \bigoplus_{j\geq 0} H_{MM}^{2j}(X,\mathbb{Q}(j))$ is a Weil cohomology for which the standard conjectures are satisfied (cf. [Kl]).

Here one has to modify the usual definition in an obvious way (like in 6.1), so that the Weil cohomology now takes values in a $\mathbb{Q}$-linear, rigid abelian tensor category.

c) Over a finite field k one expects $F^1H_M^i(X,\mathbb{Q}(j)) = H_M^i(X;\mathbb{Q}(j))_o = 0$, see §12. Over a global field k one expects $F^2H_M^i(X,\mathbb{Q}(j)) = 0$, i.e., that there are "no motivic 2-extensions" up to torsion, a statement which is implicit in Beilinson's conjectures in [Bei 1], cf. also [Bei 2] 8.5.1, and explicit in [D10] . Here the spectral sequence 11.3 should become a short exact sequence

$$0 \to R^1\Gamma_{MM}H^{i-1}_{MM}(X,\mathbb{Q}(j)) \to H^i_M(X,\mathbb{Q}(j)) \to \Gamma_{MM}H^i_{MM}(X,\mathbb{Q}(j)) \to 0 ,$$

like for the Deligne cohomology, cf. 9.7 c). Since G_k has cohomological dimension two, one has to have a vanishing or a truncation for $H^2(G_k,-)$ to get ℓ-adic cohomology close to motivic cohomology here, cf. §12 and §13.

d) As Beilinson remarks, formula 11.3.1 can be refined to

$$(11.4.1) \qquad Gr^{\nu}_F CH^j(X)\otimes\mathbb{Q} \cong R^{\nu}\Gamma((L^{j-1}H^{2j-\nu}(X))(j)) ,$$

since $CH^j(Y) = 0$ for $\dim Y < j$. Here $L^{\cdot}$ is the level filtration of a motive (called $N^{\cdot}$ in [Bei 4]), and one easily shows

$$(11.4.2) \qquad L^{j-1}H^{2j-\nu}(X) \xrightarrow{\sim} H^{2j-\nu}(X)/N^{j-\nu+1}$$

for the filtration by coniveau $N^{\cdot}$. Thus 11.1 would imply that $A_o(X)$ is representable if $H^{2d-\nu}(\bar{X},\mathbb{Q}_\ell) = N^{d-\nu+1}H^{2d-\nu}(\bar{X},\mathbb{Q}_\ell)$ for $\nu = 2,3,\dots,d = \dim X$, in generalization of Bloch's conjectures and results for $d = 2$.

<u>11.5.</u> Up to now there does not exist a definition (not even a conjectural one) of a category $\mathcal{MM}$ of mixed motives with the properties above, so one defines (absolute) motivic cohomology by K-theory and studies approximations by various realizations. These should be given by tensor categories with weights $\mathcal{T}$ (e.g., $\mathcal{T} = WRep_c(G_k,\mathbb{Q}_\ell)$ for finitely generated k , $\mathcal{T} = \mathcal{MH}$ for $k = \mathbb{C},\dots$) and complexes $\underline{R\Gamma}_{\mathcal{T},Z}(X,j)$, $\underline{R\Gamma'}_{\mathcal{T}}(X,b)$ as above together with morphisms of twisted Poincaré duality theories

$$(11.5.1) \qquad \begin{array}{ll} H^i_{\mathcal{M}}(X,\mathbb{Q}(j)) \xrightarrow{r} H^i_{\mathcal{T}}(X,j) & \text{(similar with supports)}, \\ H^{\mathcal{M}}_a(X,\mathbb{Q}(b)) \xrightarrow{r'} H^{\mathcal{T}}_a(X,b) & \end{array}$$

into the associated absolute (co-)homology (Ideally, r and r' would be induced by 11.3.2 and functors of tensor categories $\phi:\mathcal{MM} \to \mathcal{T}$ transforming the $\underline{R\Gamma}_{\mathcal{MM}}$-complexes into $\underline{R\Gamma}_{\mathcal{T}}$-complexes).

For smooth, projective X the spectral sequence

(11.5.2) $\quad E_2^{p,q} = R^p\Gamma_T H_T^q(X,j) \Rightarrow H_T^{p+q}(X,j)$

should degenerate as in 11.3 c), and we are led to study the maps

(11.5.3) $\quad Gr^{\nu}_{F(r)} H^i_M(X,\mathbb{Q}(j)) \to R^{\nu}\Gamma_T H_T^{i-\nu}(X,j)$,

where $F(r)^{\cdot} = r^{-1}F^{\cdot}$ for the filtration on $H^i_T(X,j)$ given by 11.5.2. The first two steps are given by the Chern characters

$$H^i_M(X,\mathbb{Q}(j)) \to \Gamma_T H^i_T(X,j)$$

discussed in §5 and §8, and "Abel-Jacobi maps"

$$H^i_M(X,\mathbb{Q}(j))_o \to R^1\Gamma_T H_T^{i-1}(X,j) ,$$

cf. §9. For a finite or a global field this should suffice, cf. remark 11.4 c). For fields of higher transcendence degree, however, one must also consider the maps 11.5.3 for $\nu \geq 2$, since F^2 in general does not vanish (e.g., $F^2\ CH_o(X) = T(X)$) . In particular, for $k = \mathbb{C}$ Hodge theory, and hence Deligne cohomology, is not sufficient to catch motivic cohomology, since $R^{\nu}\Gamma_H = 0$ for $\nu \geq 2$ (see 9.3 b)).

The non-vanishing of F^2 for smooth, projective X over big fields k has the following consequence for smooth varieties. By 9.10 the composition res $\circ$ r in the (conjectural) diagram

$$\begin{array}{ccc} \Gamma_{MM}H^i_{MM}(U,\mathbb{Q}(j)) & \xrightarrow{\Gamma\phi} & \Gamma_T H^i_T(U,j) \\ \text{res}\uparrow & & \uparrow\text{res} \\ H^i_M(U,\mathbb{Q}(j)) & \xrightarrow{r} & H^i_T(U,j) \end{array} \qquad (11.5.4)$$

is non-surjective for certain $U \subseteq X$ (and certain i,j). In particular (look at $T = MM$) the left restriction map will not be surjective, and the spectral sequence 11.5.2 will not degenerate for general smooth U/k. It is in this case better to treat the geometric and absolute theories separately. Note that both $\Gamma\phi$ and r could be isomorphisms; in any case we should rather regard the image of $\Gamma\phi$ as the subspace of algebraic elements in $\Gamma_T H^i_T(U,j)$.

In the rest of this section we discuss ℓ-adic and Hodge theoretic realization functors over arbitrary fields. Note that the existence of injective "regulator maps" $H^i_{\mathcal{M}}(X,\mathbb{Q}(j)) \overset{r}{\to} H^i_{\mathcal{T}}(X,j)$ together with the standard conjectures for the associated geometric theory $H^i_T(X,j)$ would imply Beilinson's conjecture 11.1, by taking the filtration $F(r)^{\cdot}$ defined above.

<u>11.6.</u> Beilinson conjectures that the ℓ-adic realization functor is faithful on $\mathcal{MM}$ - this corresponds to the injectivity of $\Gamma\phi$ - and on $D^b(M(S,\mathbb{Q}))$ for a scheme S of finite type over $\mathbb{Z}$ - this would imply that

$$H^i_{\mathcal{M}}(X,\mathbb{Q}(j)) \to H^i_{\acute{e}t}(X,\mathbb{Q}_\ell(j))$$

is injective for varieties over a finite field k, see §12 for more precise conjectures. For arbitrary fields k étale cohomology is too small (e.g., for $k = \bar{k}$), and we use the following construction. Write $X = \varprojlim_{\alpha\in A} X_\alpha$, with an inverse system $(X_\alpha)_{\alpha\in A}$ of schemes of finte type over $\mathbb{Z}[\frac{1}{\ell}]$, with affine transition maps. Then

(11.6.1) $\quad \tilde{H}^i_{\acute{e}t}(X,\mathbb{Q}_\ell(j)) := \varinjlim_{\alpha\in A} H^i_{\acute{e}t}(X_\alpha,\mathbb{Q}_\ell(j))$

is well defined and independent of the choice of $(X_\alpha)_{\alpha\in A}$ by [EGA IV] 8.13.5, and one obtains a regulator map

(11.6.2) $\quad r: H^i_{\mathcal{M}}(X,\mathbb{Q}(j)) \to \tilde{H}^i_{\acute{e}t}(X,\mathbb{Q}_\ell(j))$,

since by [Qui 1] §7, 2.2

$$K_m(X) = \varinjlim_{\alpha\in A} K_m(X_\alpha) ,$$

and this carries over to the Adams eigenspaces. Beilinson's conjecture amounts to the injectivity of 11.6.2 (in fact, one should even expect the injectivity of $r\otimes\mathbb{Q}_\ell$) for smooth X. Note that $\mathbb{Z}_\ell$ - or $\mathbb{Q}_\ell$-cohomology does not commute with inverse limits, in contrast to the case of finite coefficients.

<u>11.7. Remark</u> For extending $\tilde{H}^{\cdot}_{et}$ to a satisfactory theory as in 11.5, including a treatment of more general sheaves then $\mathbb{Q}_\ell(j)$, one may proceed as follows.

Let R_o be a ring, let R be an R_o-algebra, and let $\mathcal{R}$ be the inductive system of its subrings of finite type over R_o. Let FT/R be the category of separated schemes of finite type over R, then for $X \in ob(FT/R)$ there is an $R_1 \in \mathcal{R}$ and a scheme $\mathcal{X}_1$ of finite type over R_1 with $X = \mathcal{X}_1 \times_{R_1} R$. Let $\mathcal{X}_{R'} = \mathcal{X}_1 \times_{R_1} R'$ for $R' \in \mathcal{R}$, $R' \supseteq R_1$, and define an <u>R_o-potential ℓ-adic sheaf</u> F on X/R as an ℓ-adic sheaf $(F_{R'})_{R' \supset R}$ on the inverse system $(\mathcal{X}_{R'})_{R' \supset R_2}$ for some $R_2 \in \mathcal{R}$, $R_2 \supseteq R_1$. By using [EGA IV] 8.8.2 one defines morphisms of these as morphisms of Ind-objects and obtains an abelian category, fibred over FT/R, such that

$$H^i_{\acute{e}t}(X/R;R_o,F) = \varinjlim_{R' \supseteq R_2} H^i_{cont}(\mathcal{X}_{R'}, F_{R'})$$

only depends on X and F, in a functorial way. For $R_o = \mathbb{Z}[\frac{1}{\ell}]$ we simply talk of potential ℓ-adic sheaves and obvious have

$$\tilde{H}^i_{et}(X,\mathbb{Q}_\ell(j)) = H^i_{\acute{e}t}(X/R;\mathbb{Z},\mathbb{Q}_\ell(j)) ,$$

where the left hand side is defined as in 11.6.1 (R does not have to be a field for this). From [EGA IV] §8 one easily obtains the following properties.

a) Call an R_o-potential sheaf $F = (F_{R'})_{R' \supset R_2}$ on X/R constructible if all $F_{R'}$ are constructible and there is an $R_3 \supset R_2$ such that the transition maps $p^*F_{R'} \to F_{F''}$ for p: Spec R" → Spec R' are isomorphisms for $R' \supseteq R_3$. If k is a finitely generated field, then the constructible potential sheaves on $X \in ob(FT/k)$ can be identified with a full subcategory of the category of constructible ℓ-adic sheaves on X, by sending F to its <u>limit</u> $\varinjlim_{R'} p^*_{R'}F_{R'}$, where $p_{R'}: X \to \mathcal{X}_{R'}$ is the projection.

b) If $f: X \to Y$ is a morphism in FT/R and F is an R_o-potential ℓ-adic sheaf on X/R, then there is a spectral sequence

$$E_2^{p,q} = H^p(Y/R;R_o,R^qf_*F) \Rightarrow H^{p+q}(X/R;R_o,F) ,$$

with certain canonical R_o-potential ℓ-adic sheaves on Y/R. If R_o is a Dedekind ring, R^qf_* respects constructible R_o-potential sheaves. If $R = k$ is a finitely generated field, $R^qf_*\mathbb{Q}_\ell(j)$ can be identified with the usual sheaf $R^q f_* \mathbb{Q}_\ell(j)$ via a) (Use the generic base change theorem of [SGA 4 $\frac{1}{2}$] [finitude]).

c) $H^*_{\text{ét}}(-/R;R_o,\mathbb{Q}_\ell(*))$ is part of a twisted Poincaré duality on FT/R . (Define homology by

$$H_a^{\text{ét}}(X/R;R_o,\mathbb{Q}_\ell(b)) = \lim_{\substack{\to \\ R'\supseteq R_1}} H^{-a}_{\text{ét}}(X_\alpha, Rf^!\mathbb{Q}_\ell(-b)) ,$$

where $f_{R'}: X_{R'} \to \operatorname{Spec} R'$ is the structural morphism and the transition maps are induced by the "base change" morphisms of [SGA 4] 3.1.14.2).

For obtaining the Poincaré duality of c) as in 11.5, by $\underline{R\Gamma}$-complexes in a suitable derived category of potential ℓ-adic sheaves on Spec k one may use the techniques of [J4] . The spectral sequence 11.5.2 will be the spectral sequence

$$(11.7.1)\quad E_2^{p,q} = H^q(k/k\,;R_o,H^q(\bar{X},\mathbb{Q}_\ell(j))) \Rightarrow H^{p+q}(X/k;R_o,\mathbb{Q}_\ell(j))$$

from b) above, for $Y=$ Spec k .

<u>11.8.</u> Let us briefly discuss the Hodge realizations. As we remarked above, Deligne cohomology is expected to be a good absolute cohomology for varieties over $\mathbb{Q}$, but not for fields k with many parameters. In view of 11.6 we may try to work with

$$\tilde{H}^i_{\mathcal{D}}(X,\mathbb{Q}(j)) = \lim_{\substack{\to \\ R'}} H^i(X_{R'}\times_{\mathbb{Q}}\mathbb{C},\mathbb{Q}(j)) ,$$

$X/k = R$, $R_o = \mathbb{Q}$, R' running over the $\mathbb{Q}$-algebras of finite type in k , and $X_{R'}/R'$ as in 11.7, but it remains the question for a filtration $F^\cdot$ on $\tilde{H}^i_{\mathcal{D}}$ such that $Gr_F^\nu \tilde{H}^i_{\mathcal{D}}(X,\mathbb{Q}(j))$ only depends on the motive $H^{i-\nu}(X)$, or even a category $\mathcal{T}$ such that

$$Gr_F^\nu \tilde{H}^i_{\mathcal{D}}(X,\mathbb{Q}(j)) = Ext^\nu_{\mathcal{T}}(1,H_T^{i-\nu}(X)(j))$$

for the associated geometric realization H_T^* .

This would follow from a suitable theory of a Deligne cohomology (or absolute Hodge cohomology) for variations of Hodge structures. Namely, for X smooth and projective over k there is an $R' = R_1$ as above such that $R^q(f^{\mathbb{C}}_{R'})_*\mathbb{Q}$ is a variation of Hodge structures for $R' \supseteq R_1$, where $f^{\mathbb{C}}_{R'} : X^{\mathbb{C}}_{R'} \to S^{\mathbb{C}}_{R'}$ is the base extension to $\mathbb{C}$ (over $\mathbb{Q}$) of $f_{R'}: X_{R'} \to \operatorname{Spec} R' = S_{R'}$. The theory should give a spectral sequence

$$E_2^{p,q} = H^p_{\mathcal{D}}(S^{\mathbb{C}}_{R'}, R^q(f^{\mathbb{C}}_{R'})_*\mathbb{Q}(j)) \Rightarrow H^{p+q}_{\mathcal{D}}(X^{\mathbb{C}}_{R'}, \mathbb{Q}(j)) ,$$

which would provide the wanted spectral sequence and filtration for $\tilde{H}^i_{\mathcal{D}}(X,\mathbb{Q}(j))$ by passing to the limit over R'. The initial terms should sit in a short exact sequence

$$R^1\Gamma_H H^{p-1}(S^C_{R'}, R^q(f^C_{R'})_*\mathbb{Q}(j)) \rightarrowtail H^p_{\mathcal{D}}(S^{\mathbb{C}}_{R'}, R^q(f^{\mathbb{C}}_{R'})_*\mathbb{Q}(j)) \twoheadrightarrow \Gamma_H H^p(S^{\mathbb{C}}_{R'}, R^p(f^{\mathbb{C}}_{R'})_*\mathbb{Q}(j)) ,$$

assuming that there is a canonical mixed Hodge structure on the groups $H^r(S^{\mathbb{C}}_{R'}, R^s(f^{\mathbb{C}}_{R'})_*\mathbb{Q}(j))$ (this has not yet been established in full generality).

<u>11.9.</u> Let us point out the analogy with the ℓ-adic case. The spectral sequence 11.7.1 is obtained as the limit over the spectral sequences

$$E_2^{p,q} = \tilde{H}^p_{\text{ét}}(S_{R'}, R^q(f_{R'})_*\mathbb{Q}_\ell(j)) \Rightarrow \tilde{H}^{p+q}_{\text{ét}}(X_{R'}, \mathbb{Q}_\ell(j)) .$$

Instead of the above short exact sequence there is a spectral sequence

$$E_2^{r,s} = \tilde{H}^r_{\text{ét}}(\operatorname{Spec}\mathbb{Q}, H^s_{\text{et}}(\overline{S_{R'}}, R^q(\bar{f}_{R'})_*\mathbb{Q}_\ell(j))) \Rightarrow \tilde{H}^{r+s}_{\text{ét}}(S_{R'}, R^q(f_{R'})_*\mathbb{Q}_\ell(j)) ,$$

but a further modification of $\tilde{H}^{\nu}_{\text{ét}}$ should in fact give short exact sequences, too (cf. remark 11.4 c)). Here $\bar{f}_{R'}: \bar{X}_{R'} \to \bar{S}_{R'}$ is the base extension of $f_{R'}$ to $\bar{\mathbb{Q}}$ (over $\mathbb{Q}$).

To make the analogy even more apparent consider the case that k is finitely generated, that R' has k as field of fractions and that $\bar{S}_{R'}$ is connected and an Artin neighborhood of $\operatorname{Spec}\bar{k}$, hence an étale $K(\pi,1)$. Then

(11.9.1) $$H^s_{\text{ét}}(\bar{S}_{R'}, R^q(\bar{f}_{R'})_*\mathbb{Q}_\ell) = H^s_{\text{cont}}(\pi_1(\bar{S}_{R'}), H^q(X\times_k\bar{k}, \mathbb{Q}_\ell)) ,$$

while

(11.9.2) $$H^s(S^{\mathbb{C}}_{R'}, R^q(F^{\mathbb{C}}_{R'})_*\mathbb{Q}) = H^s(\pi_1^{top}(S^{\mathbb{C}}_{R'}), H^q(X\times_k\mathbb{C}, \mathbb{Q})) ,$$

where π_1 denotes the algebraic and π_1^{top} the topological fundamental group (with respect to the base points induced by the generic point Spec k → Spec R') . Note that $\pi_1(\bar{S}_{R'})$ is isomorphic to the profinite completion of $\pi_1^{top}(S_{R'}^{\mathbb{C}})$ and that this is compatible with the actions on $H^q(X\times_k\bar{k},\mathbb{Q}_\ell)$ and $H^q(X\times_k\mathbb{C},\mathbb{Q})$ via the comparison isomorphism.

Thus the treatment of parameters (via the $S_{R'}$) is quite parallel and produces potential $G_{\mathbb{Q}}$-representations 11.9.1 and mixed Hodge structures "over $\mathbb{Q}$" 11.9.2, and the mystery lies in the correspondence between the latter two, i.e., in the crucial case $k = \mathbb{Q}$.

Of course the solution of this mystery should be that 11.9.1 and 11.9.2 appear as realizations of the same mixed motive over $\mathbb{Q}$.

PART III

K-THEORY AND ℓ-ADIC COHOMOLOGY

§12. Finite fields and global function fields

Let k be a finite field of characteristic p, let ℓ be a prime different from p, and let X be a smooth, projective variety over k. Then one conjectures (see [Bei 4] 1.0)

12.1. Conjecture The cycle map induces an isomorphism

$$CH^j(X) \otimes \mathbb{Q}_\ell \xrightarrow{\sim} H^{2j}(\bar{X},\mathbb{Q}_\ell(j))^{G_k}$$

for all $j \geq 0$.

Obviously this is equivalent to the conjunction of the Tate conjecture B) and the conjecture that $CH^j(X)_o$, the subgroup generated by the cycles homologous to zero, is a torsion group. In fact, one should expect that $CH^j(X)_o$ is a finite group. This is known to be true for $j = 1$ (since for the abelian variety $\underline{Pic}^o_{X/k}$ the group $\underline{Pic}^o_{X/k}(k) = Pic^o(X) = CH^1(X)_o$ of k-rational points is finite) and for $j = d$, if X has pure dimension d (see [CSS] théorème 5).

By Grothendieck's isomorphism

$$K_o(X)^{(j)} \cong CH^j(X) \otimes \mathbb{Q} ,$$

conjecture 12.1 can also be reformulated in terms of K-theory, or the motivic cohomology $H^{2j}_M(X,\mathbb{Q}(j)) = K_o(X)^{(j)}$. For the other K-groups one has (see [Bei 1] 2.4.2.3)

12.2. Conjecture (Parshin) $K_m(X)$ is a torsion group for $m \neq 0$.

Again, these groups are in fact expected to be finite, since by a general conjecture of Bass they should be finitely generated. This finiteness is known to be true for dim $X \le 1$, by results of Quillen [Q2] and Harder [Har 1].

One can combine conjectures 12.1 and 12.2 into

<u>12.3 Conjecture</u> For all $i,j \in \mathbb{Z}$ one has

$$H^i_{\mathcal{M}}(X,\mathbb{Q}(j)) \otimes \mathbb{Q}_\ell \xrightarrow{\sim} H^i(\bar{X},\mathbb{Q}_\ell(j))^{G_k} .$$

Here the map is either the zero map or given by the Chern character on the algebraic K-group. To see the equivalence with 12.1 and 12.2 note that

$$H^i(\bar{X},\mathbb{Q}_\ell(j))^{G_k} = 0 \quad \text{for} \quad i \neq 2j$$

by the Weil conjectures, and that

$$K_m(X)\otimes\mathbb{Q} \cong \bigoplus_{j\ge 0} K_m(X)^{(j)} = \bigoplus_{j\ge 0} H^{2j-m}_{\mathcal{M}}(X,\mathbb{Q}(j)) ,$$

see [Bei 1] 2.2.

As in previous chapters we suggest to extend this to arbitrary varieties Z over k; for smooth varieties this goes back to Friedlander and Beilinson (see [Bei 2] 8.3.4b)).

<u>12.4 Conjecture</u> a) Let Z be a variety over k , then

$$H^{\mathcal{M}}_a(Z,\mathbb{Q}(b))\otimes\mathbb{Q}_\ell \xrightarrow{\sim} H^{ét}_a(\bar{Z},\mathbb{Q}_\ell(b))^{G_k} .$$

b) Let U be a smooth variety over k , then

$$H^i_{\mathcal{M}}(U,\mathbb{Q}(j)) \otimes \mathbb{Q}_\ell \xrightarrow{\sim} H^i_{ét}(\bar{U},\mathbb{Q}_\ell(j))^{G_k} .$$

Here the map in b) is given by the Chern character and in a) by the "Riemann-Roch transformation" as constructed by Gillet, cf. 8.3. Of course, b) follows from a) by Poincaré duality, cf. 8.1 b), and b) trivially implies conjecture 12.3. For a conclusion in the converse direction we recall the following well-known

<u>12.5. Semisimplicity Conjecture</u> (Grothendieck/Serre) The action of G_k on $H^i(\bar{X},\mathbb{Q}_\ell)$ is semi-simple for X/k smooth and proper.

More generally, one expects

<u>12.6. Semisimplicity Conjecture for arbitrary varieties</u> (cf. [D7])

a) For a variety Z over k, the action of G_k on $Gr^W_m H_a(\bar{Z},\mathbb{Q}_\ell)$ is semi-simple for all $a,m \in \mathbb{Z}$.

b) In particular, for $a,b \in \mathbb{Z}$ the eigenvalue 1 of the Frobenius endomorphism on $H_a(\bar{Z},\mathbb{Q}_\ell(b))$ is semi-simple.

<u>12.7. Theorem</u> a) If resolution of singularities holds, then conjectures 12.3 and 12.5 (that is, Grothendieck/Serre + Tate + Parshin + $CH^{\cdot}(X)_o \otimes \mathbb{Q} = 0$ for smooth, projective X) are equivalent to conjectures 12.4 and 12.6. The same holds if one only considers varieties of dimension $\leq d$.

b) Let Z be a variety over k, $Z' \subseteq Z$ a closed subvariety, and $U = Z \setminus Z'$ the open complement. If, for a fixed $b \in \mathbb{Z}$, conjectures 12.4 a) and 12.6 b) are true for two of the varieties Z, Z' and U, they are also true for the third one.

c) If 12.4 and 12.6 b) hold for all smooth varieties of dimension $\leq d$ over k, they hold for all varieties of dimension $\leq d$.

<u>Proof</u> b) Assume 12.4 and 12.6 b) for Z' and U, the other cases are similar. In the following we set $\Gamma = G_k$, $H_a(\bar{Y}) = H_a(\bar{Y},\mathbb{Q}_\ell(b))$, and $H^M_a(Y) = H^M_a(Y,\mathbb{Q}(b)) \otimes \mathbb{Q}_\ell$ for short. In the exact sequence

$$\ldots \to H_{a+1}(\bar{U}) \to H_a(\bar{Z'}) \xrightarrow{\alpha} H_a(\bar{Z}) \xrightarrow{\beta} H_a(\bar{U}) \to H_{a-1}(\bar{Z'}) \to \ldots$$
$$H_a(\bar{Z'}) \twoheadrightarrow X_a \hookrightarrow H_a(\bar{Z}) \twoheadrightarrow Y_a \hookrightarrow H_a(\bar{U})$$

let $X_a = \mathrm{Im}\,\alpha$, $Y_a = \mathrm{Im}\,\beta$ as indicated. By 12.6 b) for U and Z' the top row in the commutative diagram

$$\begin{array}{ccccccccc} 0 \to Y_a^\Gamma & \to & H_a(\bar{U})^\Gamma & \to & H_{a-1}(\overline{Z'})^\Gamma & \to & X_{a-1}^\Gamma & \to & 0 \\ \uparrow & & \uparrow\wr & & \uparrow\wr & & & & \\ H_a^M(Z) & \to & H_a^M(U) & \to & H_{a-1}^M(Z') & & & & \end{array}$$

is exact. By 12.4 for U and Z' and exactness of the bottom row the map $H_a^M(Z) \to H_a^M(\bar{Z})^\Gamma \to Y_a^\Gamma$ is surjective. This shows the exactness of

(12.7.1) $\quad 0 \to X_a^\Gamma \to H_a(\bar{Z})^\Gamma \to Y_a^\Gamma \to 0$.

Hence the top row in the commutative diagram

$$\begin{array}{ccccccccccc} \to & H_{a+1}(\bar{U})^\Gamma & \to & H_a(\overline{Z'})^\Gamma & \to & H_a(\bar{Z})^\Gamma & \to & H_a(\bar{U})^\Gamma & \to & H_{a-1}(\overline{Z'})^\Gamma & \to \\ & \uparrow\wr & & \uparrow\wr & & \uparrow & & \uparrow\wr & & \uparrow\wr & \\ & H_{a+1}^M(U) & \to & H_a^M(Z') & \to & H_a^M(Z) & \to & H_a^M(U) & \to & H_{a-1}^M(Z') & \to \end{array}$$

is exact, and we deduce the isomorphism $H_a^M(Z) \xrightarrow{\sim} H_a(\bar{Z})^\Gamma$ with the 5-lemma. Conjecture 12.6 b) for $H_a(\bar{Z})$ follows from 12.7.1 with the lemma below.

c) now follows by induction on the dimension, the case of dimension zero being trivial. Given a variety Z , the induction step consists in applying b) to a smooth, open, dense subvariety $U \subset Z$ and $Z' = Z \smallsetminus U$. Note that we may take U affine, hence quasi-projective.

a) If one has resolution of singularities, one may use simplicial varieties as in [J2] §2 to get a spectral sequence converging to $H_*(\bar{Z},\mathbb{Q}_\ell)$ and identifying $Gr_m^W H_a(\bar{Z},\mathbb{Q}_\ell)$ with a subquotient of $H_\mu(\bar{Y},\mathbb{Q}_\ell(\nu))$ for suitable $\mu,\nu \in \mathbb{Z}$ and a certain smooth and proper variety Y of dimension $\leq \dim Z$. By using Chow's lemma one may even assume that Y is projective. Thus one can deduce the semi-simplicity for Z from the conjecture of Grothendieck and Serre.

Conjectures 12.4 and 12.6 b) for arbitrary varieties can be deduced from the conjunction of 12.4 and 12.6 b) for smooth, projective varieties, again by induction on the dimension. First we treat the case of a smooth variety U, by applying b) to a smooth compactification X (resolution of singularities) and $Z' = X \smallsetminus U$. Then we get the result for arbitrary varieties by c).

<u>12.8. Lemma</u> Let k be a finite field. Call a (finite-dimensional) $\mathbb{Q}_\ell$-representation V of $\Gamma = G_k$ 1-semi-simple, if the Frobenius eigenvalue 1 is semi-simple on V. Let

$$0 \to A \to B \to C \to 0$$

be an exact sequence of $\mathbb{Q}_\ell$-representations of Γ.

a) B is 1-semi-simple if and only if A and C are 1-semi-simple and the sequence

$$(12.8.1) \qquad 0 \to A^\Gamma \to B^\Gamma \to C^\Gamma \to 0$$

is exact. In particular, if

$$\ldots \to B_{m+1} \to B_m \to B_{m-1} \to \ldots$$

is a long exact sequence of $\mathbb{Q}_\ell$-Γ-representations, with B_m 1-semi-simple for all $m \in \mathbb{Z}$, then

$$\ldots \to B_{m+1}^\Gamma \to B_m^\Gamma \to B_{m-1}^\Gamma \to \ldots$$

is exact.

b) If B has a weight filtration (i.e., is a mixed $\mathbb{Q}_\ell$-sheaf on Spec k) and $Gr_0^W B$ is semi-simple, then B is 1-semi-simple.

<u>Proof</u> a) It is clear that 1-semi-simplicity carries over to quotients and subrepresentations. If $Fr \in \Gamma$ is the Frobenius, we have an exact sequence

$$0 \to V^\Gamma \to V \xrightarrow{Fr-1} V \to V_\Gamma \to 0$$

for any representation V, so the snake lemma shows that

$$0 \to A_\Gamma \to B_\Gamma \to C_\Gamma \to 0$$

is exact if and only if 12.8.1 is. The claim now easily follows from the commutative exact diagram

$$\begin{array}{ccccccc} & & A_\Gamma & \to & B_\Gamma & \to & C_\Gamma & \to & 0 \\ & & \uparrow\alpha & & \uparrow\beta & & \uparrow\gamma & & \\ 0 & \to & A^\Gamma & \to & B^\Gamma & \to & C^\Gamma & & \end{array}$$

and the fact that V is 1-semi-simple if and only if $V^\Gamma \to V_\Gamma$ is an isomorphism.

b) One has a Γ-isomorphism $B \cong \bigoplus_{m\in\mathbb{Z}} Gr_m^W B$, and the eigenvalue 1 only appears in $Gr_o^W B$ (note that $Ext^1_\Gamma(B,B') \cong Hom_{\mathbb{Q}_\ell}(B,B')_\Gamma$).

12.9 Remark Let us single out the two principles underlying theorem 12.7:

a) The semi-simplicity conjecture implies that

$$(X,Z) \rightsquigarrow H^i_Z(\bar{X},\mathbb{Q}_\ell(j))^{G_k}$$
$$X \rightsquigarrow H_a(\bar{X},\mathbb{Q}_\ell(b))^{G_k}$$

forms a twisted Poincaré duality theory.

b) A morphism between twisted Poincaré duality theories is an isomorphism in homology if it is an isomorphism for smooth varieties. If one has resolution of singularities, the last question can be reduced to smooth and proper varieties.

In b) it actually sufficies to have a weak form of resolution of singularities: every smooth variety U has to contain an open, dense subvariety V which has a smooth compactification. Homology is better behaved than cohomology, since by the relative exact sequence 6.1 f) one can cut any variety into pieces to study its homology. For cohomology one encounters the problem that $H^i_Z(X,j)$ in general depends on the embedding of Z in X if X is singular (besides the problem that $H^i_{M,Z}(X,\mathbb{Q}(j))$ has not yet been defined for singular X).

For the application of the 5-lemma it is very important to have an isomorphism for smooth and proper varieties; if one just has an injection or a surjection it is not at all clear how this extends to arbitrary varieties. One encounters this problem, if one tries to extend Beilinson's conjectures to arbitrary varieties, since it is not clear how to extend Beilinson's groups $H^i_M(X/\mathbb{Z},\mathbb{Q}(j))$ to arbitrary varieties.

From theorem 12.7 we can deduce the following

12.10. Theorem Conjectures 12.4 and 12.6 are true for dim $Z \leq 1$ and for rational surfaces (not necessarily proper or smooth).

Proof For a smooth and proper variety X of dimension ≤ 1 Parshin's conjecture is true as remarked above, Tate's conjecture is trivially true, and the finiteness of $CH^j(X)_o$, $j = 0,1$, is also known. Furthermore, the conjecture of Grothendieck and Serre is true for H^o and H^2 (trivial) and for $H^1(\bar{X},\mathbb{Q}_\ell) = T_\ell \mathrm{Pic}(\bar{X}) \otimes_{\mathbb{Z}_\ell} \mathbb{Q}_\ell(-1)$ by Tate [T2] . The case of arbitrary curves now follows with 12.7, since one has resolution of singularities for dim $Z \leq 1$.

More explicitely, let $\tilde{Z}$ be the normalization of Z, let $U = Z^{reg}$ be the regular locus of Z, $T = Z \smallsetminus U$, and $\tilde{T} = \tilde{Z} \smallsetminus U$. Then the exaxt sequences

$$(12.10.1)\quad \begin{array}{l} \ldots \to H_a(T) \to H_a(Z) \to H_a(U) \to H_{a-1}(T) \to \ldots . \\ \ldots \to H_a(\tilde{T}) \to H_a(\tilde{Z}) \to H_a(U) \to H_{a-1}(\tilde{T}) \to \ldots , \end{array}$$

where we have set $H_a(Y) = H_a(\bar{Y},\mathbb{Q}_\ell)$ for short, show

$$Gr^W_{-1}H_1(Z) \cong Gr^W_{-1} H_1(U) \cong H_1(\tilde{Z}) ,$$

$$Gr^W_o H_1(Z) \subseteq Gr^W_o H_o(U) \subseteq H_o(\tilde{T}) .$$

Now $Gr^W_m H_1(Z) = 0$ for $m \neq 0,-1$ and

$$(12.10.2)\quad H_a(\bar{Y},\mathbb{Q}_\ell) = \begin{cases} \mathbb{Q}_\ell(d)^{C(\bar{Y})} & , \ a=d:= \dim Y , \\ \mathbb{Q}_\ell{}^{C_c(\bar{Y})} & , \ a = 0 , \\ 0 & , \ a>2d \text{ or } a < 0 , \end{cases}$$

for a variety Y , where $C(\bar{Y})$ (resp. $C_c(\bar{Y})$) is the set of irreducible (resp. compact connected) components of dimension d of Y (regarding the Galois action on these!). This shows the semi-simplicity for Z; for 12.4 and 12.6b) one may just apply 12.7b) to 12.10.1.

A rational surface Z is birationally equivalent to a variety $P = \mathbb{P}^2_k \amalg \ldots \amalg \mathbb{P}^2_k$, i.e., there is some dense open subvariety $Z \supseteq U \overset{j}{\subseteq} P$. Since all conjectures are known for P(see 14.1), and for $Y = Z \smallsetminus U$ and $Y' = P \smallsetminus U$ by the first step, we obtain 12.4 and 12.6 b) for Z by applying 12.7 b) twice. For the semi-simplicity look at the exact sequences

$$\ldots \to H_a(Y) \to H_a(Z) \to H_a(U) \to \ldots ,$$
$$\ldots \to H_a(P) \xrightarrow{j^*} H_a(U) \to H_{a-1}(Y') \to \ldots .$$

We may choose U affine and such that the map j^* is zero for a=2, since $H_2(\mathbb{P}^2_k, \mathbb{Q}_\ell(1))$ is generated by the class of a hyperplane section. Then $H_a(U) = 0$ for $a \leq 1$, and $H_a(U) \to H_{a-1}(Y')$ is injective for a = 2,3,; note that $H_3(P) = 0$. We obtain

$$Gr^W_m H_a(Y) \twoheadrightarrow Gr^W_m H_a(Z) \quad \text{for } a = 0,1, \ m \in \mathbb{Z} ,$$
$$Gr^W_m H_3(Z) \hookrightarrow Gr^W_m H_3(U) \hookrightarrow Gr^W_m H_2(Y') \quad \text{for } m \in \mathbb{Z} ,$$
$$Gr^W_m H_2(Z) \hookrightarrow Gr^W_m H_2(U) \hookrightarrow Gr^W_m H_1(Y') \quad \text{for } m = 0,-1 ,$$
$$Gr^W_{-2} H_2(Y) \twoheadrightarrow Gr^W_{-2} H_2(Z) ,$$
$$Gr^W_m H_2(Z) = 0 \quad \text{for } m \neq 0,-1,-2 ,$$

using the bounds for the weights given by the formulae in 6.5a). Together with 12.10.2 we have reduced the question to the known cases of Y and Y' .

<u>12.11. Remark</u> For a variety (resp. a smooth variety) of dimension d over k , the right hand side in conjecture 12.4 a) (resp. 12.4 b)) vanishes for $a < 2b$ or $a > b+d$ or $b < 0$ (resp. $i > 2j$ or $i < j$ or $j > d$), see 8.12 (resp. 5.11). For the left hand side the vanishing is known for $a < 2b$ or $b > d$ or $b > a$ (resp. $i > 2j$ or $j < 0$ or $i > j+d$), by results of Suslin and Soulé, see [Sou 3] Théorème 8 - where the condition "$j<-n$" has to be replaced by "$j > n$" (resp. loc. cit. proposition 5). For any field Beilinson and Soulé conjecture a vanishing for $a > 2d$ (resp. $i < 0$), see [Sou 3] 2.9 and [Bei 1] 2.2.2.

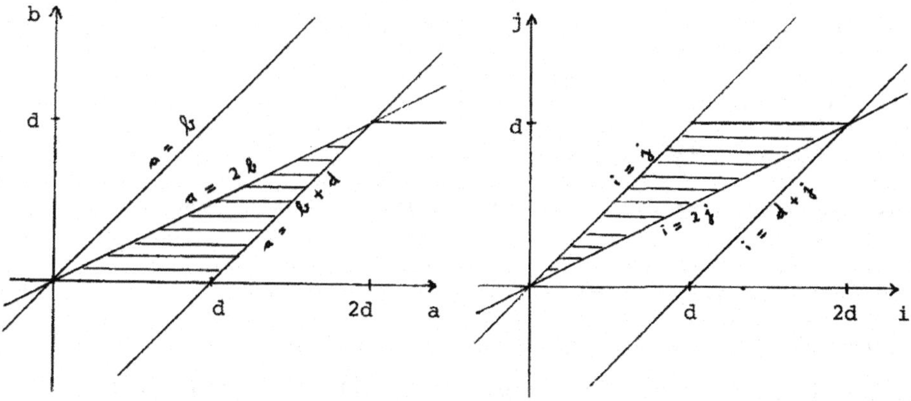

<u>12.12. Theorem</u> For a variety (resp. a smooth variety) of dimension d over the finite field k , conjecture 12.4 a) (resp. 12.4 b)) is true for $(a,b) = (2d,d)$, $(2d-1,d-1)$ or $(0,0)$ (resp. $(i,j) = (0,0)$, $(1,1)$, $(2d,d)$) .

<u>Proof</u> We have only to show the statement for 12.4 a), the other case follows by Poincaré duality. The case $(2d,d)$ is clear, in fact, the cycle class induces an isomorphism

$$Z_d(Z)\otimes_{\mathbb{Z}}\mathbb{Z}_\ell = CH_d(Z)\otimes_{\mathbb{Z}}\mathbb{Z}_\ell \xrightarrow{\sim} H_{2d}(Z,\mathbb{Z}_\ell(d)) \xrightarrow{\sim} H_{2d}(\bar{Z},\mathbb{Z}_\ell(d))^{G_k}.$$

The case $(2d-1,d-1)$ follows from 8.13.5: with the notations there we have a commutative diagram

$$\begin{array}{ccccccccc} 0 \to & H^1_{cont}(G_k,H_{2d}(\bar{Z},\mathbb{Z}_\ell(d-1))) & \to & H_{2d}(Z,\mathbb{Z}_\ell(d-1)) & \to & H_{2d-1}(\bar{Z},\mathbb{Z}_\ell(d-1))^{G_k} & \to 0 \\ & \uparrow\wr & & \uparrow\wr & & \uparrow\wr & \\ 0 \to & (\bigoplus_{x\in Z_{(d)}} k_x^\times)\otimes\mathbb{Z}_\ell & \to & E^2_{d,-d+1}(Z)\otimes\mathbb{Z}_\ell & \to & (E^2_{d,-d+1}(Z)/\bigoplus_{x\in Z_{(d)}} k_x^\times)\otimes\mathbb{Z}_\ell & \to 0, \end{array}$$

since the groups $k_x^\times$ are finite. For the same reason this induces an isomorphism

$$H^M_{2d-1}(Z,\mathbb{Q}(d-1))\otimes_{\mathbb{Q}}\mathbb{Q}_\ell \cong E^2_{d,-d+1}(Z)\otimes_{\mathbb{Z}}\mathbb{Q}_\ell \xrightarrow{\sim} H_{2d-1}(\bar{Z},\mathbb{Q}_\ell(d-1))^{G_k}$$

coinciding with the transformation τ .

The case $(a,b) = (0,0)$ follows with the spectral sequence of Bloch and Ogus

$$E^1_{p,q}(0) = \bigoplus_{x \in \bar{Z}_{(p)}} H_{p+q}(x, \mathbb{Z}_\ell(0)) \Rightarrow H_{p+q}(\bar{Z}, \mathbb{Z}_\ell(0)) ,$$

where by definition

$$H_a(x, \mathbb{Z}_\ell(b)) = \varinjlim_{\bar{U}} H_a(U, \mathbb{Z}_\ell(b))$$

the limit running over all open subvarieties $U \subseteq \overline{\{x\}}$. Now one has

$$E^2_{p,q}(b) = 0 \quad \text{for} \quad q < 0$$

as follows easily from the smooth case by excision for the E^2-terms (cf. [Gi] §8), where it holds since $\mathcal{H}^q(\mathbb{Z}_\ell(j))$, the sheaf associated to $\bar{U} \rightsquigarrow H^q(\bar{U}, \mathbb{Z}_\ell(j))$, vanishes for $q > d$ by ([Mi] VI 7.1). Therefore $E^2_{0,0}(0)$ is isomorphic to $H_o(\bar{Z}, \mathbb{Z}_\ell(0))$. On the other hand, Bloch and Ogus show that

$$E^2_{0,0}(0) = Z_o(\bar{Z})/N_1Z_o(\bar{Z})$$

with $N_rZ_p(\bar{Z}) = \{\alpha \in Z_p(\bar{Z}) \mid$ there exists $Y \in Z_{p+r}(\bar{Z})$ such that $\mathrm{Supp}(\alpha) \subseteq \mathrm{Supp}(Z)$ and $cl(\alpha) = 0$ in $H_{2p}(Y, \mathbb{Z}_\ell(p))\}$ (see [BO] 7.2, where in fact the index $p+r$ in (7.2.2) has to be replaced by $p+r-1$). I claim that the image of $N_1Z_o(\bar{Z})$ in $CH_o(\bar{Z})$ is torsion. Obviously this has to be checked for $d = 1$, i.e., for a curve. But then it follows, e.g., from 12.10. We obtain an isomorphism

$$CH_o(\bar{Z}) \otimes \mathbb{Q}_\ell \xrightarrow{\sim} H_o(\bar{Z}, \mathbb{Q}_\ell) ,$$

which implies the claim, since

$$CH_o(Z) \otimes \mathbb{Q}_\ell \xrightarrow{\sim} (CH_o(\bar{Z}) \otimes \mathbb{Q}_\ell)^{G_k} .$$

12.13 Now let k be a global function field, that is, an algebraic function field in one variable over a finite field $\mathbb{F}_q$. Without restriction we may assume that $\mathbb{F}_q$ is algebraically closed in k . Then there exists a smooth, projective, geometrically connected curve C over $\mathbb{F}_q$ with function field k .

If Z is a variety over k , then there exists an open subvariety $W \subseteq C$ and a flat model Z_W of Z over W (that is

$f: Z_W \to W$ flat, of finite type, with $Z_W \times_W \mathrm{Spec}\, k \cong Z$). Two such models become isomorphic over a suitable open subvariety W' (see [EGA IV] 8.8.2.5).

12.14. Definition Let $\ell \neq p = \mathrm{char}\, k$ be a prime and let $k_\infty = k \cdot \bar{\mathbb{F}}_q$. Call a (finite-dimensional, continuous) $\mathbb{Q}_\ell$-representation V of G_k arithmetic, if it comes from a smooth $\mathbb{Q}_\ell$-sheaf over some W as above (that is, if $G_k \to \mathrm{Aut}(V)$ factorizes through $\pi_1(W, \mathrm{Spec}\, \bar{k})$). In this case let

$$\tilde{H}^\nu(k,V) = \varinjlim_U H^\nu_{cont}(\pi_1(U,\mathrm{Spec}\, \bar{k}),V) ,$$
$$\tilde{H}^\nu(k_\infty,V) = \varinjlim_{\bar{U}} H^\nu_{cont}(\pi_1(\bar{U},\mathrm{Spec}\, \bar{k}_\infty),V) ,$$

where the inductive limit is over all open $U \subseteq W$, and $\bar{U} = U \times_{\mathbb{F}_q} \bar{\mathbb{F}}_q$.

12.15. Remark In the notation of 11.7, V is just a potential sheaf on Spec k, and

$$\tilde{H}^\nu(k,V) = H^\nu(\mathrm{Spec}\, k/k;\mathbb{F}_q,V) ,$$
$$\tilde{H}^\nu(k_\infty,V) = H^\nu(\mathrm{Spec}\, k_\infty/k_\infty;\bar{\mathbb{F}}_q,V) .$$

12.16. Theorem Let Z be smooth over k, and for $U \subseteq W$ open, with the notations of 12.13, let $Z_U = Z_W \times_W U = f^{-1}(U)$.

a) If conjecture 12.4 is true for (i,j) and Z_U, for U running through a cofinal system of open subvarieties of W, then there is an exact sequence

$$0 \to \tilde{H}^1(k_\infty, H^{i-1}(\bar{Z},\mathbb{Q}_\ell(j)))^\Gamma \to H^i_M(Z,\mathbb{Q}(j)) \otimes \mathbb{Q}_\ell \xrightarrow{r_{i,j}} H^i(\bar{Z},\mathbb{Q}_\ell(j))^{G_k},$$

with $\Gamma = \mathrm{Gal}(k_\infty/k) \cong \mathrm{Gal}(\bar{\mathbb{F}}_q/\mathbb{F}_q)$.

b) Let $U \subseteq W$ be open such that Z_U is smooth and $R^i f_* \mathbb{Q}_\ell$ is smooth over U. If

$$H^i_M(Z_U,\mathbb{Q}(j)) \otimes \mathbb{Q}_\ell \to H^i_{\acute{e}t}(\bar{Z}_U,\mathbb{Q}_\ell(j))^\Gamma$$

is surjective and $H^i_{\acute{e}t}(\bar{Z}_U,\mathbb{Q}_\ell(j))$ is 1-semi-simple (12.8), for $\bar{Z}_U = Z_U \times_{\mathbb{F}_q} \bar{\mathbb{F}}_q$, then

$$r_{i,j}: H^i_M(Z,\mathbb{Q}(j)) \otimes \mathbb{Q}_\ell \to H^i_{ét}(\bar{Z},\mathbb{Q}_\ell(j))^{G_k}$$

is surjective.

c) If for some U as in b) and some closed point $x \in U$

$$(H^i_M(Z_x,\mathbb{Q}(j))/H^i_M(Z_x,\mathbb{Q}(j))_o)\otimes\mathbb{Q}_\ell \to H^i_{ét}(\bar{Z}_x,\mathbb{Q}_\ell(j))$$

is injective, $Z_x = Z_U\times_U\kappa(x)$ the fibre of f over x and $\bar{Z}_x = Z_x\times_{\mathbb{F}_q}\bar{\mathbb{F}}_q$, then the same is true for

$$r_{i,j}:(H^i_M(Z,\mathbb{Q}(j))/H^i_M(Z,\mathbb{Q}(j))_o) \otimes \mathbb{Q}_\ell \to H^i_{ét}(\bar{Z},\mathbb{Q}_\ell(j)) ,$$

where we set $H^i_M(Z,\mathbb{Q}(j))_o = \mathrm{Ker}(r_{i,j}:H^i_M(Z,\mathbb{Q}(j)) \to H^i_{ét}(\bar{Z},\mathbb{Q}_\ell(j))^{G_k})$, similarly for Z_x .

d) In particular, if conjectures 12.4 and 12.6 b) are true for one Z_U and all fibres Z_x , $x \in U$ closed, then

$$r_{i,j}: \quad (H^i_M(Z,\mathbb{Q}(j))/H^i_M(Z,\mathbb{Q}(j))_o)\otimes\mathbb{Q}_\ell \xrightarrow{\sim} H^i_{ét}(\bar{Z},\mathbb{Q}_\ell(j))^{G_k} \qquad \text{and}$$

$$\tilde{r}_{i,j}: \quad H^i_M(Z,\mathbb{Q}(j))_o \otimes \mathbb{Q}_\ell \xrightarrow{\sim} \tilde{H}^1(k_\infty,H^{i-1}_{ét}(\bar{Z},\mathbb{Q}_\ell(j)))^\Gamma$$

are isomorphisms for all $i,j \in \mathbb{Z}$, and all Tate conjectures are true for Z .

<u>Proof</u> a), b): We may assume that U is affine; then $cd_\ell(\bar{U}) \le 1$ (see [Mi] VI 7.2) and hence the Leray spectral sequence for f gives an exact sequence

$$0\to H^1(\bar{U},R^{i-1}f_*\mathbb{Q}_\ell(j))\to H^i(\bar{Z}_U,\mathbb{Q}_\ell(j)) \to H^0(\bar{U},R^if_*\mathbb{Q}_\ell(j)) \to 0 .$$

We obtain an exact sequence

$$0\to H^1(\bar{U},R^{i-1}f_*\mathbb{Q}_\ell(j))^\Gamma\to H^i(\bar{Z}_U,\mathbb{Q}_\ell(j))^\Gamma \to H^0(\bar{U},R^if_*\mathbb{Q}_\ell(j))^\Gamma$$

(12.16.1)

$$\begin{array}{c} H^i(\bar{Z}_U,\mathbb{Q}_\ell(j))^\Gamma \\ \uparrow \mathrm{ch} \\ H^i_M(Z_U,\mathbb{Q}(j)) \end{array}$$

with the Chern character as indicated, where the right vertical map is surjective if $H^i(\bar{Z}_U,\mathbb{Q}_\ell(j))$ is 1-semi-simple.

By Deligne's generic base change theorem ([SGA 4 $\frac{1}{2}$][finitude]) the sheaves $R^{\nu}f_* \mathbb{Q}_\ell$ are smooth if U is small enough; then they can be identified with the $\pi_1(\bar{U},\mathrm{Spec}\ \bar{k}) =: \pi_1(\bar{U})$-representations $(R^{\nu}f_*\mathbb{Q}_\ell)_{\mathrm{Spec}\ \bar{k}} = H^{\nu}(\bar{Z},\mathbb{Q}_\ell)$ and the diagram 12.16.1 with the following one (cf. [Mi] V 2.17)

$$(12.16.2)\qquad \begin{array}{ccccc} 0\to H^1_{cont}(\pi_1(\bar{U}),H^{i-1}(\bar{Z},\mathbb{Q}_\ell(j)))^{\Gamma} & \to & H^i(\bar{Z}_U,\mathbb{Q}_\ell(j))^{\Gamma} & \to & H^i(\bar{Z},\mathbb{Q}_\ell(j))^{\pi_1(U)} \\ & & \uparrow \mathrm{ch} & & \\ & & H^i_M(Z_U,\mathbb{Q}(j)) & & . \end{array}$$

By passing to the limit over the U we get a commutative exact diagram

$$(12.16.3)\qquad \begin{array}{ccccc} 0\to \tilde{H}^1(k_\infty,H^{i-1}(\bar{Z},\mathbb{Q}_\ell(j)))^{\Gamma} & \to & \varinjlim_{U} H^i(\bar{Z}_U,\mathbb{Q}_\ell(j))^{\Gamma} & \to & H^i(\bar{Z},\mathbb{Q}_\ell(j))^{G_k} \\ \uparrow \tilde{r}_{i,j}\otimes\mathbb{Q}_\ell & & \uparrow \widetilde{\mathrm{ch}}\otimes\mathbb{Q}_\ell & \nearrow\ r_{i,j}\otimes\mathbb{Q}_\ell & \uparrow \\ 0\to H^i_M(Z,\mathbb{Q}(j))_0\otimes\mathbb{Q}_\ell & \to & H^i_M(Z,\mathbb{Q}(j))\otimes\mathbb{Q}_\ell & \to & (H^i_M(Z,\mathbb{Q}(j))/H^i_M(Z,\mathbb{Q}(j))_0)\otimes\mathbb{Q}_\ell \to 0 , \end{array}$$

using that

$$(12.16.4)\qquad K_{2j-i}(Z)^{(j)} = \varinjlim_{U} K_{2j-i}(Z_U)^{(j)}$$

cf. [Q1] §7, 2.2. The assumption of a) implies that $\widetilde{\mathrm{ch}}\otimes\mathbb{Q}_\ell$ is an isomorphism, so the statements of a) and b) are clear.

For c) we use the commutative diagram

$$(12.16.5)\qquad \begin{array}{ccc} H^i_M(Z_x,\mathbb{Q}(j)) & \xrightarrow{r_{i,j}} & H^i_{\acute{e}t}(\bar{Z}_x,\mathbb{Q}_\ell(j)) \\ sp_M\uparrow & & \uparrow sp \\ H^i_M(Z,\mathbb{Q}(j)) & \xrightarrow{r_{i,j}} & H^i_{\acute{e}t}(\bar{Z},\mathbb{Q}_\ell(j)) , \end{array}$$

where sp is the specialization map in étale cohomology - which can be obtained as the dual of the generization map for cohomology with compact support, cf. [DV] exp. 0 , 4.4 - and where sp_M is the speciali zation map in motivic cohomology. The latter one can be defined as follows (cf. Gillet's construction and remark on K' in [Gi] 8.6 ff.): for $V \subseteq U$ open, sufficiently small, let $t \in \mathcal{O}(V)^{\times}$ be a local parameter at $x \in U$, and also denote by t the corresponding element in

$H^1_M(V,\mathbb{Q}(1)) = \mathcal{O}(V)^\times \otimes \mathbb{Q}$ (cf. 6.12.4 e)). Then t maps to the fundamental class of x under δ in the exact sequence

$$\ldots \to H^1_M(U,\mathbb{Q}(1)) \to H^1_M(V,\mathbb{Q}(1)) \xrightarrow{\delta} H^2_{M,Z}(U,\mathbb{Q}(1)) \to H^2_M(U,\mathbb{Q}(1))$$

$Z = U \setminus V$, and one defines a map

$$sp_t : H^i_M(Z_V,\mathbb{Q}(j)) \xrightarrow{Uf^*(t)} H^{i+1}_M(Z_V,\mathbb{Q}(j+1)) \xrightarrow{\delta} H^i_M(Z_x,\mathbb{Q}(j)) \ ,$$

where δ comes from the exact sequence

$$\ldots \to H^{i+1}_M(Z_U,\mathbb{Q}(j+1)) \to H^{i+1}_M(Z_V,\mathbb{Q}(j+1)) \xrightarrow{\delta} H^{i+2}_{M,Z_U \setminus Z_V}(Z_U,\mathbb{Q}(j+1)) \to \ldots$$

$$H^{i+2}_{M,Z_U \setminus Z_V}(Z_U,\mathbb{Q}(j+1)) \cong \bigoplus_{x \in U \setminus V} H^i_M(Z_x,\mathbb{Q}(j))$$

This depends on the choice of t, but the induced map sp_M on

$$H^i_M(Z,\mathbb{Q}(j)) = \varinjlim_V H^i_M(Z_V,\mathbb{Q}(j))$$

does not. The same construction can be carried out in étale cohomology, with t replaced by its first Chern class $t_{ét} \in H^1_{ét}(\bar{V},\mathbb{Q}_\ell(1))$, and for the commutativity of 12.16.5 it remains to show that

$$\begin{array}{ccc} H^i_{ét}(\bar{Z}_U,\mathbb{Q}_\ell(j)) & \xrightarrow{\delta\circ(-Uf^*(t_{ét}))} & H^i_{ét}(\bar{Z}_x,\mathbb{Q}_\ell(j)) \\ & \searrow \qquad \nearrow sp & \\ & H^i_{ét}(\bar{Z},\mathbb{Q}_\ell(j)) & \end{array}$$

commutes. This follows from the compatibility of specialization with cupproducts, with similar arguments as in [DV] exp. VI, 3.6.

Now c) is clear from the commutative diagram

$$\begin{array}{ccc} H^i_M(Z_x,\mathbb{Q}(j))/H^i_M(Z_x,\mathbb{Q}(j))_o & \hookrightarrow & H^i_{ét}(\bar{Z}_x,\mathbb{Q}_\ell(j)) \\ \uparrow & & \wr\uparrow sp \\ H^i_M(Z,\mathbb{Q}(j))/H^i_M(Z,\mathbb{Q}(j))_o & \hookrightarrow & H^i_{ét}(\bar{Z},\mathbb{Q}_\ell(j)) \ , \end{array}$$

in which sp is an isomorphism by the smoothness of $R^i f_* \mathbb{Q}_\ell$.

d) is now clear from a)-c) and 12.7 b): by induction it follows that conjectures 12.4 and 12.6b) hold for all Z_V , $V \subset U$, and if in the diagram 12.16.3 $\widetilde{ch} \otimes \mathbb{Q}_\ell$ is bijective, then $\tilde{r}_{i,j} \otimes \mathbb{Q}_\ell$ is an isomorphism if the right vertical map is injective.

<u>12.17. Remarks</u> a) A similar theorem holds for arbitrary varieties

and homology. The specialization map for étale homology is the dual of the generization map for cohomology with compact support, and for the motivic homology one has to use the pull-back morphism f^* for the flat map f(cf. 14.4 below). Formula 12.16.4 becomes

$$H_a^{M}(Z,\mathbb{Q}(b)) = \varinjlim_{U} H_{a+2}^{M}(Z_U,\mathbb{Q}(b+1))$$

and follows from [Q1] §7 (2.4), and everything can be proved by considering cohomology with supports.

b) If Z is smooth and proper, then

$$H^i(\bar{Z},\mathbb{Q}_\ell(j))^{G_k} = 0 \qquad \text{for } i \neq 2j ,$$

$$\tilde{H}^1(k_\infty,H^{i-1}(\bar{Z},\mathbb{Q}_\ell(j)))^\Gamma = \begin{cases} H^1_{cont}(G_k,H^{i-1}(\bar{Z},\mathbb{Q}_\ell(j)) , & i<2j-1 \text{ or } i = 2j \\ \tilde{H}^1(k,H^{i-1}(\bar{Z},\mathbb{Q}_\ell(j))) , & i = 2j-1, \\ 0 , & i > 2j . \end{cases}$$

In fact, one has a Hochschild-Serre spectral sequence

$$0 \to H^1_{cont}(\Gamma,H^{i-1}(\bar{Z},\mathbb{Q}_\ell(j))^{G_{k_\infty}}) \to \tilde{H}^1(k,H^{i-1}(\bar{Z},\mathbb{Q}_\ell(j))) \xrightarrow{res} \tilde{H}^1(k_\infty,H^{i-1}(\bar{Z},\mathbb{Q}_\ell(j)))^\Gamma \to 0 ,$$

and $H^r(\bar{Z},\mathbb{Q}_\ell(j))$ is pure of weight $r-2j$, implying the first claim and the vanishing of

$$H^1_{cont}(\Gamma,H^{i-1}(\bar{Z},\mathbb{Q}_\ell(j))^{G_{k_\infty}}) = (H^{i-1}(\bar{Z},\mathbb{Q}_\ell(j))^{G_{k_\infty}})_\Gamma$$

for $i-1 \neq 2j$. The map

$$\tilde{H}^1(k,H^{i-1}(\bar{Z},\mathbb{Q}_\ell(j))) \xrightarrow{inf} H^1_{cont}(G_k,H^{i-1}(\bar{Z},\mathbb{Q}_\ell(j)))$$

is injective, and an isomorphism for $i \neq 2j-1$, see [J3] lemma 4 . Finally, for U as in 12.16 b), $H^1_{cont}(\pi_1(\bar{U}) , H^{i-1}(\bar{Z},\mathbb{Q}_\ell(j))) = H^1(\bar{U},R^{i-1}f_*\mathbb{Q}_\ell(j))$ is mixed of weights $\geq i-2j$ ([D9] 3.3.5), so

$$\tilde{H}^1(k_\infty,H^{i-1}(\bar{Z},\mathbb{Q}_\ell(j)))^\Gamma = 0 \quad \text{for } i > 2j .$$

The above investigations suggest the following conjecture, which sharpens and extends the conjectures 8.5 and 9.15 (for a global function field k).

12.18. Conjecture a) Let Z be a variety over k, and for $a,b \in \mathbb{Z}$ let $H_a^M(Z,\mathbb{Q}(b))_o = \operatorname{Ker}(r'_{a,b}: H_a^M(Z,\mathbb{Q}(b)) \to H_a(\bar{Z},\mathbb{Q}(b))^{G_k})$ (cf. 9.4). Then

$$(H_a^M(Z,\mathbb{Q}(b))/H_a^M(Z,\mathbb{Q}(b))) \otimes \mathbb{Q}_\ell \to H_a(\bar{Z},\mathbb{Q}_\ell(b))^{G_k}$$

$$\tilde{r}'_{a,b} \otimes \mathbb{Q}_\ell : H_a^M(Z,\mathbb{Q}(b))_o \otimes \mathbb{Q}_\ell \to \tilde{H}^1(k_\infty, H_{a+1}^M(\bar{Z},\mathbb{Q}_\ell(b)))^\Gamma$$

are isomorphisms.

b) In particular, for X smooth and proper over k,

$$\widetilde{ch}_{i,j} \otimes \mathbb{Q}_\ell : H_M^i(X,\mathbb{Q}(j)) \otimes \mathbb{Q}_\ell \to \tilde{H}^1(k, H^{i-1}(\bar{X},\mathbb{Q}_\ell(j)))$$

is an isomorphism for $i < 2j$, and the Abel-Jacobi map

$$cl' \otimes \mathbb{Q}_\ell : CH^j(X)_o \otimes \mathbb{Q}_\ell \to H^1_{cont}(G_k, H^{2j-1}(\bar{X},\mathbb{Q}_\ell(j)))$$

is an isomorphism for $j \geq 0$.

12.19 Remark Formally, part b) can be expressed for all $i,j \in \mathbb{Z}$ by saying that

$$H_M^i(X,\mathbb{Q}(j))_o \otimes \mathbb{Q}_\ell \to \tilde{H}^1(k, W_{-1}H^{i-1}(\bar{X},\mathbb{Q}_\ell(j)))$$

should be an isomorphism for all $i,j \in \mathbb{Z}$. If $H^{i-1}(\bar{X},\mathbb{Q}_\ell(j))$ is semi-simple, one has

$$\tilde{H}^1(k, W_{-1}H^{i-1}(\bar{X},\mathbb{Q}_\ell(j))) = \operatorname{Ext}^1_{S_a^{s.s}(k,\mathbb{Q}_\ell)}(\mathbb{Q}_\ell, H^{i-1}(\bar{X},\mathbb{Q}_\ell(j))),$$

where $S_a^{s.s}(k,\mathbb{Q}_\ell)$ is the category of arithmetical $\mathbb{Q}_\ell$-sheaves F with weight filtration on Spec k, for which the $Gr_m^W F$ are semi-simple. The motivic interpretation of this (compare §11) would be that the ℓ-adic realization functor is faithful and that there are no motivic 2-extensions over k (up to torsion).

12.20. Theorem a) Conjecture 12.18 is true for smooth varieties X and $(a,b) = (2d-1,d-1)$ or $(2d,d)$, $d = \dim X$.

b) Conjecture 12.18 is true for $X = \operatorname{Spec} k$.

Proof a) This follows from 12.16.3 for $(i,j) = (1,1), (0,0)$: By 12.12 the middle vertical map is an isomorphism, and for (i,j)

$= (1,1)$ the right vertical map is an isomorphism by 5.16.

b) This again follows from 12.16.3. With the notations there we may take $Z_U = U$, and by 12.10 the middle vertical map is an isomorphism. If $i \neq 0,1$, or if $i = 0$ and $j \neq 0$, then all groups vanish. The case $(i,j) = (0,0)$ has been treated above. For $i = 1$ we have $H^1(\mathrm{Spec}\ \bar{k},\mathbb{Q}_\ell(j)) = 0$, hence $H^1_M(\mathrm{Spec}\ k,\mathbb{Q}(j)) = H^1_M(\mathrm{Spec}\ k,\mathbb{Q}(j))_o)$ and the claim follows.

§13. Number fields

In analogy with the conjecture 12.18b) for global function fields we conjecture the following.

13.1. Conjecture Let k be a finite extension of $\mathbb{Q}$, let X be a smooth projective variety over k, and let ℓ be a prime. Then the map

$$\tilde{r}_{i,j}: H^i_M(X,\mathbb{Q}(j))\otimes_{\mathbb{Q}}\mathbb{Q}_\ell \to H^1_{cont}(G_k,H^{i-1}(\bar{X},\mathbb{Q}_\ell(j)))$$

induced by the Chern character is an isomorphism for $i < j$ and injective for $i = j$, $(i,j) \neq (0,0)$.

For further discussion and motivation we refer the reader to [J3] §2, where this conjecture is stated for $i < j$. If in analogy with 12.14 we define

$$\begin{aligned}\tilde{H}^\nu(k,H^i(\bar{X},\mathbb{Q}_\ell(j))) &= H^\nu(k/k;\mathbb{Z},H^i(\bar{X},\mathbb{Q}_\ell(j)))\\ &= \varinjlim_U H^\nu_{cont}(\pi_1(U),H^i(\bar{X},\mathbb{Q}_\ell(j))),\end{aligned} \tag{13.1.1}$$

where U runs over all open subschemes of $\mathrm{Spec}\ \mathcal{O}_k \smallsetminus S$, with $\mathcal{O}_k$ the ring of integers and S a finite set of primes of k containing all those above ℓ or where X has bad reduction, then we have

$$\tilde{H}^1(k,H^i(\bar{X},\mathbb{Q}_\ell(j))) \xrightarrow[\sim]{\inf} H^1_{cont}(G_k,H^i(\bar{X},\mathbb{Q}_\ell(j))) \quad \text{for } i \neq 2j-1,$$

see [J3] lemma 4. In that paper we also discuss the following

<u>13.2. Conjecture</u> For X/k as above one has

$$\tilde{H}^2(k,H^i(\bar{X},\mathbb{Q}_\ell(j))) = 0 \quad \text{for} \quad i+1 < j \ .$$

In view of the spectral sequence

$$E_2^{p,q} = \tilde{H}^p(k,H^q(\bar{X},\mathbb{Q}_\ell(j))) \Rightarrow \tilde{H}^{p+q}(X,\mathbb{Q}_\ell(j))$$

(cf.11.7.1) and the fact that

$$H^q(\bar{X},\mathbb{Q}_\ell(j))^{G_k} = 0 \quad \text{for} \quad q \neq 2j \ ,$$

conjecture 13.2 is equivalent to

<u>13.2'. Conjecture</u> For X as above one has

$$\tilde{H}^i(X,\mathbb{Q}_\ell(j)) \xrightarrow{\sim} \tilde{H}^1(k,H^{i-1}(\bar{X},\mathbb{Q}_\ell(j)))$$

for $i \leq j$, $(i,j) \neq (0,0)$.

Hence we may combine 13.1 and 13.2 to

<u>13.3. Conjecture</u> For X as above the map

$$H^i_M(X,\mathbb{Q}(j)) \otimes_{\mathbb{Q}} \mathbb{Q}_\ell \to \tilde{H}^i(X,\mathbb{Q}_\ell(j))$$

induced by the Chern character (cf.11.6.2) is an isomorphism for $i < j$ and injective for $i = j$.

We want to extend this to arbitrary varieties.

<u>13.4. Theorem</u> a) If 13.2 is true for smooth, projective varieties of dimension $\leq d$ then for any variety X of dimension d over k

$$\tilde{H}^2(k,H_a(\bar{X},\mathbb{Q}_\ell(b))) = 0 \quad \text{for} \quad a > d+b+1 \ .$$

b) If 13.3 is true for smooth, projective varieties of dimension $\leq d$, then for any variety X of dimension d over k the map

$$H_a^M(X,\mathbb{Q}(b)) \otimes_{\mathbb{Q}} \mathbb{Q}_\ell \to H_a(X/k,\mathbb{Q}_\ell(b)) =: \tilde{H}_a(X,\mathbb{Q}_\ell(b))$$

(cf. 11.7.c)) is an isomorphism for $a > d+b$ and injective for $a = d+b$.

For the proof we use the following lemma.

13.5 Lemma Let X be a variety of dimension d over k, let $Y \subset X$ be a closed subvariety such that $U = X \setminus Y$ is dense in X.

a) If $\tilde{H}^2(k, Gr_m^W H_a(\bar{Z}, \mathbb{Q}_\ell(b))) = 0$ for $a > d+b+1$ and $Z = Y$ and one of the varieties X and U, then this vanishing also holds for the other one.

b) If the map in 13.4 b) has the property stated there for Y and one of the varieties X and U, this property also holds for the other one.

Proof a) We have a long exact sequence

$$\ldots \to Gr_m^W H_a(\bar{Y}, \mathbb{Q}_\ell(b)) \to Gr_m^W H_a(\bar{X}, \mathbb{Q}_\ell(b)) \to Gr_m^W H_a(\bar{U}, \mathbb{Q}_\ell(b)) \to Gr_m^W H_{a-1}(\bar{Y}, \mathbb{Q}_\ell(b)),$$

and since this belongs to a sequence of polarizable motives for absolute Hodge cycles (cf. 6.11.1), it can be split into a series of split short exact sequences. This shows that the sequence remains exact after applying the functor $H^2(\tilde{k}, -)$. This implies the claim, since $a > d+b+1$ implies $a-1 > \dim Y + b+1$.

b) We have a commutative diagram of exact sequences

$$\begin{array}{ccccccc}
\ldots \to H_a(Y/k, \mathbb{Q}_\ell(b)) & \to & H_a(X/k, \mathbb{Q}_\ell(b)) & \to & H_a(U/k, \mathbb{Q}_\ell(b)) & \to & H_{a-1}(Y/k, \mathbb{Q}_\ell(b)) \to \ldots \\
\uparrow & & \uparrow & & \uparrow & & \uparrow \\
\ldots \to H_a^M(Y, \mathbb{Q}(b)) \otimes \mathbb{Q}_\ell & \to & H_a^M(X, \mathbb{Q}(b)) \otimes \mathbb{Q}_\ell & \to & H_a^M(U, \mathbb{Q}(b)) \otimes \mathbb{Q}_\ell & \to & H_{a-1}^M(Y, \mathbb{Q}(b)) \otimes \mathbb{Q}_\ell \to
\end{array}$$

so the claim easily follows with the weak four-lemma (see [ML] I. 3.1) and the fact that $a > d+b$ (resp. $a = d+b$) implies $a-1 > \dim Y + b$ and $a-1 \geq d+b$ (resp. $a > \dim Y + b$ and $a-1 \geq \dim Y+b$).

Proof of 13.4: With induction on d, the induction claim for 13.4 a) being that

$$\tilde{H}^2(k, Gr_m^W H_a(\bar{X}, \mathbb{Q}_\ell(b))) = 0$$

for every $m \in \mathbb{Z}$ provided $a > d+b+1$. The case $d = 0$ is trivial, and for $d > 0$ we first show the claims for smooth quasi-projective varieties U . Choosing a smooth, projective compactification $X \supseteq U$ and letting $Y = X \setminus U$, the induction step is given by 13.5, since the claims coincide with 13.2 and 13.3, respectively, for X , by Poincaré duality and purity. For an arbitrary variety X we choose a dense, open smooth quasi-projective subvariety U for which we get the result by the first step. The induction from $Y = X \setminus U$ to X now again follows with 13.5.

13.6. Remark By the Hochschild-Serre spectral sequence for homology, statements a) and b) of theorem 13.4 together would imply that the map

$$H_a^M(X,\mathbb{Q}(b)) \otimes_{\mathbb{Q}} \mathbb{Q}_\ell \to \tilde{H}^1(k, H_{a+1}(\bar{X},\mathbb{Q}_\ell(b)))$$

is an isomorphism for $a > d+b$ and injective for $a = d+b$, $(a,b) \neq (2d,d)$. Conjecture 12.18 a) would imply exactly the same in the function field case.

13.7. At the end of this section, let us discuss the relation with a general conjecture of Beilinson on K-theory and ℓ-adic cohomology ([Bei 4] 5.10 D)vi), but note the misprint in the definition of $H^{\cdot}_{fine}(S,\mathbb{Z}/\ell^n(i))$ in the last reference, where $R\pi_*\mathbb{Z}/\ell^n(i)$ should be replaced by $\tau_{\leq i}R\pi_*\mathbb{Z}/\ell^n(i))$.

For a regular scheme X let $\alpha: X_{ét} \to X_{Zar}$ be the canonical map from the étale to the Zariski site of X, and let ℓ be a prime which is invertible on X . Beilinson notes that the complexes of Zariski sheaves

$$\Delta/\ell^n(j) := \tau_{\leq j}R\alpha_*\mathbb{Z}/\ell^n(j)$$

satisfy Gillet's axioms for a twisted Poincaré duality theory provided Grothendieck's purity conjecture (s. [SGA 5] I 3.1.4) is true. Assuming this one gets Chern classes from $K_*(X)$ into

$$H^i_{fine}(X,\mathbb{Z}/\ell^n(j)) := H^i_{Zar}(X,\Delta/\ell^n(j))$$

factorizing the Chern classes into $H^i_{\acute{e}t}(X,\mathbb{Z}/\ell^n(j))$ via the canonical map induced by

$$\tau_{\leq j}R\alpha_*\mathbb{Z}/\ell^n(j) \to R\alpha_*\mathbb{Z}/\ell^n(j) .$$

Beilinson conjectures that $H^*_{fine}(X,\mathbb{Z}/\ell^n(*))$ is the "motivic cohomology with $\mathbb{Z}/\ell^n$-coefficients". The meaning of this statement can be expressed in three (related) ways.

1) Beilinson conjectures the existence of complexes of Zariski sheaves $\mathbb{Z}_{M}(j)$ on X satisfying certain axioms (including Gillet's ones) such that via the Chern characters

(13.7.1) $$H^i_{Zar}(X,\mathbb{Z}_{M}(j)) \otimes \mathbb{Q} \cong H^i_{M}(X,\mathbb{Q}(j))$$

i.e., $H^i_{Zar}(X,\mathbb{Z}_{M}(j))$ can be regarded as an integral motivic cohomology. The above conjecture then claims that there are canonical quasi-isomorphisms

$$\Delta/\ell^n(j) \simeq \mathrm{Cone}(\mathbb{Z}_{M}(j) \xrightarrow{\ell^n} \mathbb{Z}_{M}(j)) .$$

In particular this would give long exact sequences

(13.7.2) $$\ldots \to H^i_{Zar}(X,\mathbb{Z}_{M}(j)) \xrightarrow{\ell^n} H^i_{Zar}(X,\mathbb{Z}_{M}(j)) \to H^i_{fine}(X,\mathbb{Z}/\ell^n(j)) \to \ldots .$$

2) For a variety X over a field k Bloch and Landsburg defined higher Chow groups $CH^*(X,*)$, and in [Bl5] Bloch gives much evidence for the conjecture that for a smooth variety

(13.7.3) $$H^i_{M}(X,\mathbb{Z}(j)) := CH^j(X,2j-i)$$

is the correct version of integral motivic cohomology. Bloch in particular proves that

(13.7.4) $$H^i_{M}(X,\mathbb{Z}(j)) \otimes_{\mathbb{Z}} \mathbb{Q} \cong H^i_{M}(X,\mathbb{Q}(j))$$

and that there is a cycle map from the groups 13.7.3 into any reasonable cohomology theory [Bl 6] . By the same arguments as in [Bl 6] one gets a cycle map for regular X

$$H^i_{M}(X,\mathbb{Z}/\ell^n(j)) := CH^j(X;\mathbb{Z}/\ell^n;2j-i) \to H^i_{fine}(X,\mathbb{Z}/\ell^n(j))$$

and Beilinson's conjecture can be stated more concretely as the conjecture that this map is an isomorphism for all $n,i,j \in \mathbb{Z}$

and all smooth varieties X, cf. the statements in [MS 2] .

3) One can show that one actually gets Chern classes

$$K_{2j-i}(X,\mathbb{Z}/\ell^n)^{(j)} \to H^i_{fine}(X,\mathbb{Z}/\ell^n(j)) ,$$

and Beilinson's conjecture claims that this map is an isomorphism up to "small standard factorials" (those necessary to define the Chern character).

Grothendieck's purity conjecture is unproved yet, but one can use Thomason's result that the purity conjecture is "almost" true for reasonable schemes [Th] . It implies that for schemes X of finite type over $\mathbb{Z}$

$$H^i_{fine}(X,\mathbb{Q}_\ell(j)) := (\varprojlim_n H^i_{fine}(X,\mathbb{Z}/\ell^n(j))) \otimes_{\mathbb{Z}_\ell} \mathbb{Q}_\ell$$

is part of a twisted Poincaré duality theory and that there are Chern characters

$$K_{2j-i}(X) \to H^i_{fine}(X,\mathbb{Q}_\ell(j)) .$$

Together with Bass' conjecture on the finite generation of the K-groups or similar conjectures on the finite generation of integral motivic cohomology, Beilinson's conjecture (in any of the three formulations) would imply that the induced map

$$H^i_{\mathcal{M}}(X,\mathbb{Q}(j)) \otimes_{\mathbb{Q}} \mathbb{Q}_\ell \to H^i_{fine}(X,\mathbb{Q}_\ell(j))$$

is an isomorphism for all $i,j \in \mathbb{Z}$ and X regular, of finite type over $\mathbb{Z}$. For a smooth variety X over a field k this would imply isomorphisms

$$H^i_{\mathcal{M}}(X,\mathbb{Q}(j)) \otimes_{\mathbb{Q}} \mathbb{Q}_\ell \to \tilde{H}^i_{fine}(X , \mathbb{Q}_\ell(j)) ,$$

where $\tilde{H}^i_{fine}(X , \mathbb{Q}_\ell(j))$ is defined as in 11.6 , as $\varinjlim_\alpha H^i_{fine}(X_\alpha,\mathbb{Q}_\ell(j))$, for $X = \varprojlim_\alpha X_\alpha$, with X_α of finite type over $\mathbb{Z}$.

Now the exact triangle

$$\tau_{\leq j} R\alpha_* \mathbb{Z}/\ell^n(j) \to R\alpha_* \mathbb{Z}/\ell^n(j) \to \tau_{>j} R\alpha_* \mathbb{Z}/\ell^n(j) \to$$

gives rise to a long exact sequence

$$\cdots \to H^{i-1}_{Zar}(X,\tau_{>j}R\alpha_*\mathbb{Z}/\ell^n(j)) \to H^i_{Zar}(X,\tau_{\leq j}R\alpha_*\mathbb{Z}/\ell^n(j)) \to$$

$$\| \text{ if } i\leq j+1 \qquad\qquad \|$$

$$0 \qquad\qquad H^i_{fine}(X,\mathbb{Z}/\ell^n(j))$$

$$\to H^i_{Zar}(X,R\alpha_*\mathbb{Z}/\ell^n(j)) \to H^i_{Zar}(X,\tau_{>j}R\alpha_*\mathbb{Z}/\ell^n(j)) \to \cdots$$

$$\| \qquad\qquad \| \text{ if } i\leq j$$

$$H^i_{et}(X,\mathbb{Z}/\ell^n(j)) \qquad\qquad 0 \quad .$$

Hence the map

$$H^i_{fine}(X,\mathbb{Z}/\ell^n(j)) \to H^i_{ét}(X,\mathbb{Z}/\ell^n(j))$$

is an isomorphism for $i \leq j$ and injective for $i = j+1$. Concluding, Beilinson's conjecture suggests the following one, sharpening 13.3.

<u>13.8. Conjecture</u> If X is a smooth variety over a field k, and if $\ell \neq \text{char } k$ is a prime, then

$$H^i_M(X,\mathbb{Q}(j)) \otimes_{\mathbb{Q}} \mathbb{Q}_\ell \to \tilde{H}^i(X, \mathbb{Q}_\ell(j))$$

is an isomorphism for $i \leq j$ and injective for $i = j+1$.

<u>13.9.</u> The relation between $H^i_{fine}(X,\mathbb{Z}/\ell^n(j))$ and $H^i_{et}(X,\mathbb{Z}/\ell^n(j))$ is highly non-trivial for $i > j$. Since the hypercohomology spectral sequence for $R\alpha_*\mathbb{Z}/\ell^n(j)$ is just the spectral sequence for the coniveau filtration $N^\cdot$([BO]6.4 and footnote), the truncation $\tau_{\leq j}$ in a certain way means to force the condition

(13.9.1) $\qquad H^i(X,j) = N^{i-j}H^i(X,j)$

upon the considered twisted Poincaré duality theory. This should be compared with the remarks in 5.24: condition 13.9.1 is a necessary condition to get a cohomology theory close to motivic cohomology. The various conjectures I made for finite or global fields try to calculate H^*_{fine}, i.e., the coniveau spectral sequence, in terms of the ℓ-adic realizations - compare the Tate conjecture and its generalization due to Grothendieck. In fact, the case of cycles and $i = 2j$ is the extreme case, where Beilinson's conjecture gives no information at all.

To see this, let $\mathcal{H}^i(\mathbb{Z}/\ell^n(j))$ be the Zariksi sheaf associated to $U \mapsto H^i_{et}(U,\mathbb{Z}/\ell^n(j))$, then we have (assuming purity)

$$H^p_{Zar}(X,\mathcal{H}^q(\mathbb{Z}/\ell^n(j))) = 0, \qquad p > q,$$
$$H^p_{Zar}(X,\mathcal{H}^p(\mathbb{Z}/\ell^n(p))) = CH^p(X)/\ell^n$$

by (the proof of) [BO] 7.7, and hence

$$H^i_{fine}(X,\mathbb{Z}/\ell^n(j)) = 0 \quad \text{for } i > 2j,$$
$$H^{2j}_{fine}(X,\mathbb{Z}/\ell^n(j)) \cong CH^j(X)/\ell^n$$

as remarked by Beilinson in [Bei 4]. Consequently, Beilinson's conjecture (formulation 2)) is "trivially" true for $i = 2j$, and here the mystery lies in the map

$$H^{2j}_{fine}(X,\mathbb{Z}/\ell^n(j)) \to H^{2j}_{ét}(X,\mathbb{Z}/\ell^n(j)).$$

<u>13.10 Remarks</u> a) If one assumes purity, then one has a Bloch-Ogus spectral sequence for étale cohomology (this is sometimes referred to as Gersten's conjecture for étale cohomology). From this spectral sequence one easily deduces that Beilinson's conjecture (formulation 2)) is equivalent to having (cf. [MS2])

$$CH^j(F;\mathbb{Z}/\ell^n,2j-i) \xrightarrow{\sim} H^i(F,\mathbb{Z}/\ell^n(j)), \quad i \leq j,$$

for all fields F/k (Note that trivially $CH^i(F;\mathbb{Z}/\ell^n,j) = 0 = H^i_{fine}(F,\mathbb{Z}/\ell^n(j))$ for $i > j$). In this sense, the conjecture is of "arithmetic" nature, while the determination of $H^i_{fine}(X,\mathbb{Z}/\ell^n(j)) \to H^i(X,\mathbb{Z}/\ell^n(j))$ for $j < i \leq 2j$ - related to the coniveau filtration - is of "geometric" nature, cf. the picture in 5.24.

b) Lichtenbaum has conjectured the existence of certain complexes of sheaves $\Gamma(j)$ for the étale topology with certain relations to algebraic K-theory [Li 1] and constructed candidates for $j \leq 2$ [Li 2]. By the triangle axiom (loc. cit.) and the implied sequence

$$\ldots \to H^i_{et}(X,\Gamma(j)) \xrightarrow{\ell^n} H^i_{et}(X,\Gamma(j)) \to H^i_{et}(X,\mathbb{Z}/\ell^n(j)) \to \ldots$$

the groups $H^i_{et}(X,\Gamma(j))$ are rather related to étale cohomology (i.e., can be thought of as "étale cohomology with $\mathbb{Z}$-coefficients") and are quite different from motivic cohomology. But in compatibility with Beilinson's conjecture Lichtenbaum conjectures quasi-isomorphisms

$$\mathbb{Z}_{\mathcal{M}}(j) \simeq \tau_{\leq j} R\alpha_* \Gamma(j)$$

and shows that this together with his "Hilbert 90" axiom would in fact imply Beilinson's conjecture (formulation 1)).

<u>13.11. Theorem</u> The conjectures stated in this chapter are true for $X = \operatorname{Spec} k$.

<u>Proof</u> One has $\tilde{H}^i(k,\mathbb{Q}_\ell(n)) = 0$ for $i \geq 3$,

$$H^i(\operatorname{Spec} \bar{k},\mathbb{Q}_\ell(j)) = \begin{cases} \mathbb{Q}_\ell(j) & , i = 0 , \\ 0 & , i \neq 0 , \end{cases}$$

and it follows from results of Borel and Soulé that

$$K_{2n}(k) \otimes \mathbb{Q} = 0 \quad , \quad n \geq 1 ,$$

$$K_{2n-1}(k)^{(n)} \otimes_{\mathbb{Q}} \mathbb{Q}_\ell \xrightarrow{\sim} K_{2n-1}(k) \otimes_{\mathbb{Z}} \mathbb{Q}_\ell \xrightarrow[\sim]{ch_{1,n}} \tilde{H}^1(k,\mathbb{Q}_\ell(n)) \quad , \quad n \geq 1 ,$$

$$\tilde{H}^2(k,\mathbb{Q}_\ell(n)) = 0 \qquad \text{for } n > 1 ,$$

cf. [J3] , example 3. Hence conjecture 13.2 is true for Spec k, and

$$H^0_{\mathcal{M}}(\operatorname{Spec} k,\mathbb{Q}(0)) \otimes \mathbb{Q}_\ell \overset{r_{0,0}}{\xrightarrow{\sim}} H^0(\operatorname{Spec} \bar{k},\mathbb{Q}_\ell)^{G_k} \cong \mathbb{Q}_\ell \quad ,$$

$$H^1_{\mathcal{M}}(\operatorname{Spec} k,\mathbb{Q}(j)) \otimes \mathbb{Q}_\ell \overset{r_{1,j}}{\xrightarrow{\sim}} \tilde{H}^1(k,\mathbb{Q}_\ell(j)) \quad , \quad j \geq 1 ,$$

$$H^i_{\mathcal{M}}(\operatorname{Spec} k,\mathbb{Q}(j)) = 0 \ , \ \text{otherwise},$$

showing conjecture 13.8.

§14. Linear varieties

In this section we study certain varieties, whose cohomology groups are successive extensions of Tate objects, and prove most of the conjectures stated in this paper for them. We start with two general observations.

14.1. Lemma For every conjecture stated in this paper the following principle holds: If the conjecture is true for a smooth variety X, it also holds for every affine or projective fibre bundle over X.

14.2. This is due to the fact that every considered Poincaré duality theory satisfies the following axioms for a smooth variety X (cf. [Gi] 1.2 (ix),(x), and 8.9 above):

t) (homotopy invariance) If E is a vector bundle on X and $p: V(E) \to X$ is the associated affine fibre bundle, then the morphisms

$$p^*: H^i(X,j) \to H^i(V(E),j)$$

are isomorphism for all $i,j \in \mathbb{Z}$.

n') (projective bundle isomorphism) If E is a vector bundle of rank n on X and $p: P(E) \to X$ is the associated projective fibre bundle, then

$$\bigoplus_{\nu=0}^{n} H_{a+2\nu-2n}(X,b+\nu-n) \xrightarrow{\oplus p^*\cap\xi^\nu} H_a(P(E),b)$$

is an isomorphism for all $a,b \in \mathbb{Z}$, where $\xi \in \Gamma H^2(X,1)$ is the Chern class of the canonical line bundle $\mathcal{O}(1)$ on $P = P(E)$.

14.3. Here the pull-back p^* is defined via Poincaré duality, i.e., as the arrow making the diagram

$$\begin{array}{ccc}
H_{a-2n}(X,b-n) & \longrightarrow & H_a(P,b) \\
\eta_X\cap \Big\uparrow \wr & & \wr\Big\uparrow \eta_P\cap \\
H^{2d+2n-a}(X,d+n-b) & \xrightarrow{p^*} & H^{2d+2n-a}(P,d+n-a)
\end{array} \quad , \ d=\dim X \ ,$$

commutative, and ξ^ν is the ν-fold cupproduct of ξ, where the cupproduct is defined by Poincaré duality, too, by commutativity of

$$\begin{array}{ccc}
H^i(X,j) \otimes H^{i'}(X,j') & \xrightarrow{\cup} & H^{i+i'}(X,j+j') \\
\eta_X\cap \Big\uparrow \wr \qquad \Big\| & & \wr\Big\uparrow \eta_X\cap \\
H_{2d-i}(X,d-j) \otimes H^{i'}(X,j') & \xrightarrow{\cap} & H_{2d-i-i'}(X,d-j-j') \ .
\end{array}$$

Proof of 14.1. The axioms t) and n') immediately give the claim for the semi-simplicity conjectures in §12 and for Tate's conjecture C), since $L(V(r),s)=L(V,r+s)$. All other conjectures concerned morphisms between twisted Poincaré duality theories, their bijectivity, injectivity or surjectivity, and obviously these morphisms respect the above isomorphisms. The axioms t' and n') are well-known for the ℓ-adic theory and then follow for absolute Hodge theory, and in particular the Hodge and the deRham theory, by the comparison isomorphisms. For the motivic (co-)homology t) and n') follow from Quillen's corresponding results for the K-theory ([Qui 1] §7,4), since the maps respect the Adams eigenspaces.

14.4. Remark The above can be extended to singular varieties by introducing the following concepts: i) a pull-back morphism in homology $f^*: H_a(Y,b) \to H_{a+2n}(X,b+n)$ for flat morphisms $f: X \to Y$ of fibre dimension $\leq n$,

ii) a cupproduct in cohomology, both compatible with Poincaré duality as in 14.3. For the considered Poincaré duality theories these exist, and instead of t) on has:

t') If E is a vector bundle of rank n on a variety X and p:

$V(E) \to X$ is the associated fibre bundle, then

$$p^*: H_a(X,b) \to H_{a+2n}(V(E),b+n)$$

is an isomorphism for all $a,b \in \mathbb{Z}$.

Axiom n') holds literally for arbitrary varieties X, with the pull-back and cupproduct mentioned above (See [Qui 1] loc. cit. for K-theory, which carries over to motivic homology by the methods of [Sou 3], and [DV] VIII 5 for ℓ-adic homology).

The considered Chern characters and Riemann-Roch transformations are compatible with this, hence one obtains 14.1 for arbitrary varieties, since all conjectures were formulated in terms of homology.

14.5. Lemma For every conjecture stated in this paper the following principle holds. If, for a variety X over a field k, the conjecture holds for $X\times_k K$, for a finite Galois extension K/k, then it holds for X itself.

Proof All conjectures involved functors with rational Galois descent, i.e., we have canonically

(14.5.1) $\quad H_a(X,b)\otimes_{\mathbb{Z}}\mathbb{Q} = (H_a(X\times_k K,b)\otimes_{\mathbb{Z}}\mathbb{Q})^G$,

for $G = \mathrm{Gal}(K/k)$. Here we regard $X\times_k K$ as a variety over k so that we have an action of G on $X\times_k K$ over k and an induced one on the homology. The property 14.5.1 then follows from the fact that for the functors

$$H_a(X\times_k K,b) \underset{p^*}{\overset{p_*}{\rightleftarrows}} H_a(X,b)$$

induced by the étale projection $p: X\times_k K \to X$ we have

(14.5.2)
$$p_*p^* = [K:k]\cdot \mathrm{id}\ ,$$
$$p^*p_* = \sum_{\sigma\in G}\sigma\ .$$

We make the following inductive definition.

<u>14.6. Definition</u> Let S be a scheme. Call a flat S-scheme Z 0-linear, if it is empty or isomorphic to the affine S-space $\mathbb{A}_S^N$ for some $N \geq 0$. Call Z n-linear, for $n \geq 1$, if there is a tripel $\{U,X,Y\}$ of flat S-schemes such that $Y \subseteq X$ is a closed S-immersion and $U \subseteq X$ is the open complement, Y and one of $\{U,X\}$ is $(n-1)$-linear, and Z is the other member in $\{U,X\}$. Call Z linear, if it is n-linear for some $n \geq 0$ (and note that n-linear implies $(n+1)$-linear).

Examples are S-schemes that are stratified by affine S-spaces, e.g., $Z = \mathbb{P}_S^N$. Over a field k, examples of linear k-varieties are complements in $\mathbb{P}_k^N$ (or $\mathbb{A}_k^N$) of a union of linear subspaces, or successive blow-ups of $\mathbb{P}_k^N$ in linear subspaces, Grassmannians, flag varieties, and varieties stratified by such varieties.

<u>14.7. Theorem</u> a) If X is a linear variety over a finitely generated field k and $\ell \neq \mathrm{char}(k)$, then

$$Gr_m^W H_a^{\acute{e}t}(\bar{X},\mathbb{Q}_\ell) = \begin{cases} 0 & , \quad m \text{ odd} \\ \text{sum of } \mathbb{Q}_\ell(-\nu)\text{'s}, & m = 2\nu \;, \quad \nu \in \mathbb{Z}, \end{cases}$$

as a G_k-module, for all $m,a \in \mathbb{Z}$. If $\mathrm{char}(k) = 0$, then

$$Gr_m^W H_a^{AH}(X) \otimes_{\mathbb{Z}} \mathbb{Q} = \begin{cases} 0 & , \quad m \text{ odd}, \\ \text{sum of } 1(-\nu)\text{'s} \;, & m = 2\nu \;, \; \nu \in \mathbb{Z}, \end{cases}$$

for the realization for absolute Hodge cycles (cf. 6.11) for all $a,m \in \mathbb{Z}$. In particular, the corresponding statement holds for the Hodge structures $H_a(X(\mathbb{C}),\mathbb{Q})$, if X is a linear variety over $\mathbb{C}$.

b) The Hodge conjecture and the Tate conjectures are true for linear varieties.

c) If X is a linear variety over a finite field k, then

$$r_{a,b}: H_a^{\mathcal{M}}(X,\mathbb{Q}(b)) \otimes \mathbb{Q}_\ell \xrightarrow{\sim} H_a(\bar{X},\mathbb{Q}_\ell(b))^{G_k}$$

is an isomorphism for $\ell \neq \mathrm{char}(k)$ and all $a,b \in \mathbb{Z}$. Hence in view of a) and b), all stated conjectures are true for X. Conjectures 12.4 and 12.6 b) are more generally true for linear C-varieties. C a curve over k.

d) If X is a linear variety over a global function field k, then

$$r_{a,b}:(H_a^{\mathcal{M}}(X,\mathbb{Q}(b))/H_a^{\mathcal{M}}(X,\mathbb{Q}(b))_o)\otimes\mathbb{Q}_\ell \xrightarrow{\sim} H_a^{\text{ét}}(\bar{X},\mathbb{Q}_\ell(b))^{G_k} \quad \text{and}$$

$$\tilde{r}_{a,b}: \quad H_a^{\mathcal{M}}(X,\mathbb{Q}(b))_o \otimes \mathbb{Q}_\ell \xrightarrow{\sim} \tilde{H}^1(k_\infty,H_{a+1}^{\text{ét}}(\bar{X},\mathbb{Q}_\ell(b)))$$

(notations as in 12.18) are isomorphisms for $\ell \neq \mathrm{char}(k)$ and all $a,b \in \mathbb{Z}$. Hence all stated conjectures are true for X.

e) If X is a linear variety over a number field k, then all conjectures stated in §13 are true for X, namely,

$$r_{a,b}:H_a^{\mathcal{M}}(X,\mathbb{Q}(b)) \otimes \mathbb{Q}_\ell \rightarrow \tilde{H}_a^{\text{ét}}(X,\mathbb{Q}_\ell(b))$$

is an isomorphism for $a \geq \dim X + b$ and injective for $a = \dim X + b-1$, and $\tilde{H}^2(k,H_a^{\text{ét}}(\bar{X},\mathbb{Q}_\ell(b))) = 0$ for $a > \dim X + b + 1$.

<u>Proof</u> a) It follows from the relative homology sequence (6.1 f)) and the inductive definition of linear varieties that $Gr_m^W H_a^{\text{ét}}(\bar{X},\mathbb{Q}_\ell)$, or $Gr_m^W H_a^{AH}(X)_{\mathbb{Q}}$ for $\mathrm{char}(k) = 0$, have a filtration such that the graded terms have the wanted property. For $\mathrm{char}(k) = 0$ we may use the polarizations for absolute Hodge cycles, like in lemma 1.1, to see that $Gr_m^W H_a^{AH}(X)_{\mathbb{Q}}$ is in fact a direct sum of these graded terms, and hence get the result. For $\mathrm{char}(k) = p > 0$ we have to proceed differently. Again it suffices to show that $Gr_m^W H_a^{\text{ét}}(\bar{X}, \mathbb{Q}_\ell)$ is semi-simple. For this we transport Falting's arguments in [FW] VI §3 to our setting. There exists a smooth, geometrically irreducible variety U over a finite field $\mathbb{F}_q$, with generic point $\eta = \mathrm{Spec}\ k$, such that

i) X/k extends to a linear U-scheme $f: \mathcal{X} \rightarrow U$ (this follows from the inductive definition and [EGA IV] §8),

ii) $R^{-a}f_*Rf^!\mathbb{Q}_\ell$ is smooth over U, commutes with arbitrary base change, and has a weight filtration (compare the proof of 6.8.2). One gets an exact sequence of fundamental groups

$$1 \to \pi_1(U\times_{\mathbb{F}_q}\bar{\mathbb{F}}_q) \to \pi_1(U) \to \mathrm{Gal}(\bar{\mathbb{F}}_q/\mathbb{F}_q) \to 1$$

(omitting the base points in the notation), and by Faltings' arguments in loc. cit. it suffices to show that $\pi_1(U\times_{\mathbb{F}_q}\bar{\mathbb{F}}_q)$ and the decomposition group D_x of a closed point x of U act semi-simply on $Gr^W_m H_a^{ét}(\bar{X},\mathbb{Q}_\ell)$. For $\pi_1(U\times_{\mathbb{F}_q}\bar{\mathbb{F}}_q)$ this is known by a fundamental result of Deligne [D9] 3.4.1 iii). On the other hand, the base change property in ii) above gives an isomorphism of $D_x \cong \mathrm{Gal}(\bar{\mathbb{F}}_q/\kappa(x))$-representations

$$Gr^W_m H_a^{ét}(\bar{X},\mathbb{Q}_\ell) \cong Gr^W_m H_a^{ét}(\bar{X}_x,\mathbb{Q}_\ell)\ ,$$

where $X_x/\kappa(x)$ is the fibre of X/U over $x \in U$ and $\bar{X}_x = X_x\times_{\kappa(x)}\bar{\mathbb{F}}_q$, $\kappa(x)$ the finite residue field of x. By i) X_x is a linear variety over $\kappa(x)$, so we have reduced the question to the case of a finite field, which is treated in c).

b) By a) the Hodge conjecture in this case means that

$$cl_i\otimes\mathbb{Q}\colon CH_i(X)\otimes\mathbb{Q} \to W_0H_{2i}(X,\mathbb{Q}(i))$$

is surjective for all $i \geq 0$. By the commutative exact diagram

$$\begin{array}{ccccccc}
W_0H_{2i}(Y,\mathbb{Q}(i)) & \to & W_0H_{2i}(X,\mathbb{Q}(i)) & \to & W_0H_{2i}(U,\mathbb{Q}(i)) & \to & 0 \\
\uparrow cl_i & & \uparrow cl_i & & \uparrow cl_i & & \\
CH_i(Y) & \to & CH_i(X) & \to & CH_i(U) & \to & 0\ ,
\end{array}$$

for $Y \subseteq X$ closed and $U = X\smallsetminus Y$ (compare 7.5), and the inductive definition 15.6, we reduce to the case of an affine space $\mathbb{A}^N_{\mathbb{C}}$, which is known by 14.1. The Tate conjecture A) is similar, and the finite generation of $A_i(X)$ and injectivity of $A_i(X)\otimes\mathbb{Q}_\ell \xrightarrow{cl_i\otimes\mathbb{Q}_\ell} H_{2i}^{ét}(\bar{X},\mathbb{Q}_\ell(i))$ follows from the case of a finite field via the commutative diagram

$$\begin{array}{ccc}
A_i(X) & \hookrightarrow & H_{2i}^{ét}(\bar{X},\mathbb{Q}_\ell(i)) \\
sp\downarrow & & \downarrow\wr \\
A_i(X_x) & \hookrightarrow & H_{2i}^{ét}(\bar{X}_x,\mathbb{Q}_\ell(i))
\end{array}$$

where X_x is as in a) and sp is the specialization map (compare the proof of 12.16). For a finite field k one shows in fact by induction starting from $\mathbb{A}_k^N$, that the groups $K'_m(X)$ and $E^2_{p,q}(X)$ (from Quillen's spectral sequence 6.12.5) are finitely generated for a linear variety X/k (use that these are the homology groups of a Poincaré duality theory, cf. [Gi] for the $E^2_{p,q}$). In particular, this is true for $E^2_{i,-i}(X) = CH_i(X)$, and a fortiori for $A_i(X)$. The injectivity of $cl_i \otimes \mathbb{Q}_\ell$ is a special case of c).

For Tate's conjecture C), in view of a) and the proved conjecture B), we have to show that

$$\mathrm{ord}_{s=\dim_a(k)} L(\mathbb{Q}_\ell, s) = 1 ,$$

i.e., the Tate conjecture C) for X = Spec k (Note that $L(V(i),s) = L(V,i+s)$). This is well-known for finite or global fields, where we can replace $L(\mathbb{Q}_\ell,s)$ by the zeta function $\zeta_k(s)$ of k, and follows in general from Grothendieck's formula recalled in the proof of 7.17. If F is the prime field of k, it implies that $L(\mathbb{Q}_\ell,s) = f(s)\cdot\zeta_F(s-\mathrm{tr.deg}(k))$ with f(s) holomorphic and non-vanishing at $s = \dim_a k$.

c) Since the ℓ-adic cohomology of a linear variety X/k is a successive extension of representations $\mathbb{Q}_\ell(n)$, conjectures 12.6 a) and 12.6 b) are equivalent in this case. If C is a variety over k with $\dim C \leq 1$, then conjectures 12.4 and 12. 6 b) hold for $\mathbb{A}_C^N$ by 12.10 and 14.1 (14.3 for singular C), hence for arbitrary linear C-varieties by 12.7 b) and induction.

d) follows from this by applying theorem 12.16 and remark 12.17 a) to the linear scheme X/U satisfying properties i) and ii) above, U now being a curve over a finite field. Note that we have an isomorphism

$$H_a^{\mathcal{M}}(X_x,\mathbb{Q}(b)) \otimes \mathbb{Q}_\ell \xrightarrow{\sim} H_a^{\text{ét}}(\bar{X}_x,\mathbb{Q}_\ell(b))^{\Gamma}$$

for every closed $x \in U$ by c).

e) By 13.11, 14.1 and induction with the 5-lemma (cf remark 12.9) we get that

$r_{a,b}: H_a^{\mathcal{M}}(X,\mathbb{Q}(b)) \otimes \mathbb{Q}_\ell \xrightarrow{\sim} \tilde{H}_a^{fine}(X,\mathbb{Q}_\ell(b))$

for all $a,b \in \mathbb{Z}$, where $H_\cdot^{fine}$ is the homology theory associated to $H^\cdot_{fine}$, and $\tilde{H}_\cdot^{fine}$ is obtained from it as in 11.7. For a quasi-projective scheme X over Spec $\mathbb{Z}$, $H_\cdot^{fine}$ may be defined by embedding X into a smooth scheme M of pure fibre dimension N over Spec $\mathbb{Z}$ and setting

$$H_a^{fine}(X,\mathbb{Q}_\ell(b)) = H^{2N-a}_{Zar,X}(M,\tau_{\leq N-b}R\alpha_*\mathbb{Q}_\ell(N-b)) ,$$

for $\alpha: M_{ét} \to M_{Zar}$. The $\mathbb{Q}_\ell$-purity [Th] then shows:

$$H_a^{fine}(X,\mathbb{Q}_\ell(b)) \to H_a^{ét}(X,\mathbb{Q}_\ell(b))$$

is an isomorphism for $a \geq d+b$ and injective for $a = d+b-1$, d the relative dimension of X over $\mathbb{Z}$. The claimed vanishing of $\hat{H}^2(k,H_a(\bar{X},\mathbb{Q}_\ell(b)))$ follows from 13.11, 14.1 and induction with 13.5 a), or from 13.11, 6.5 a), and a) above.

One may deduce the statement on the map from $H_a^{\mathcal{M}}$ to $\tilde{H}_a^{ét}$ also without refering to $\tilde{H}_a^{fine}$, by using 13.11, 14.1 and induction as in the proof of 13.5 b), replacing d by d-1 in the argument.

Appendix A: A letter from Bloch to the author

2/11/87

Dear Jannsen,

The homological formulation of the Hodge conjecture for singular varieties which you gave in your talk is the only one possible. In fact, some years ago Mumford suggested to me that one should look for a counterexample to the corresponding cohomological conjecture. Here is one.

Let $S_0 \subset \mathbf{P}^3$ be a smooth hypersurface of degree $d \geq 4$ defined over $\overline{\mathbf{Q}}$. Let $x \in S_0(\mathbf{C})$ be $\overline{\mathbf{Q}}$-generic and let $S/\mathbf{C}$ be the blowup of S_0 at x. Let $P = BL_{\mathbf{P}^3}(x)$, so $S \subset P$. Let $W = P \amalg_S P$. Since $H^3(S) = (0)$, we have an exact sequence of Hodge structures

$$0 \to H^4(W, \mathbf{Q}(2)) \to H^4(P, \mathbf{Q}(2))^{\oplus 2} \to H^4(S, \mathbf{Q}(2)) \to 0.$$

Note $H^2(P, \mathbf{Q}(1))$ has generators e = class of exceptional divisor and h = pullback of hyperplane from $\mathbf{P}^3$. We have

$$CH^2(P)_{\mathbf{Q}} \xrightarrow{\sim} H^4(P, \mathbf{Q}(2)) \cong \mathbf{Q} \cdot h^2 \oplus \mathbf{Q} \cdot e^2 \quad (e \cdot h = 0).$$

Also $H^4(S, \mathbf{Q}(2)) \cong \mathbf{Q}$ with $h^2 \cdot S = d$, $e^2 \cdot S = -1$. It follows that $H^4(W, \mathbf{Q}(2)) \cong \mathbf{Q}^{\oplus 3}$ (as a Hodge structure). In particular $((d-1)h^2,\ d(h^2 + e^2)) \in H^4(P, \mathbf{Q}(2))^{\oplus 2}$ comes from a Hodge class on W. But on the level of Chow groups, this class restricts to $d(x) - h^2 \cdot S \in CH_0(S) \cong CH_0(S_{0\mathbf{C}})$. Because x is $\overline{\mathbf{Q}}$-generic and $p_g(S_0) > 0$, this class is of infinite order. Thus $\mathrm{Ker}(CH^2(P)_{\mathbf{Q}}^{\oplus 2} \to CH^2(S)) \cong \mathbf{Q}^{\oplus 2}$.

Remarks 1. This discussion shows that no contravariant Chow group can provide the extra element needed.

2. I guess the Hodge conjecture is true for divisors on complex projective varieties because one has the exponential. This presupposes, however, that $gr^0_{\mathrm{Hodge}} H^2(X, \mathbf{C}) \hookrightarrow H^2(X, \mathcal{O}_X)$ always. Is this o.k. ?

3. Note the counterexample is for curves on a 3-fold, where the classical Hodge conjecture is true!

4. With a bit more work, one can get a hypersurface example. Namely take $S_0 = \mathbf{P}^3 \cap T_0 \subset \mathbf{P}^4$ where T_0 is a smooth 3-dim. hypersurface defined over $\overline{\mathbf{Q}}$. Let

$$Q = BL_{\mathbf{P}^4}(\{x\}), \quad T = BL_{T_0}(\{x\}), \quad P = BL_{\mathbf{P}^3}(\{x\}), \quad S = BL_{S_0}(\{x\})$$

so we have a cartesian square of strict transforms

$$\begin{array}{ccc} S & \hookrightarrow & T \\ \downarrow & & \downarrow \\ P & \hookrightarrow & Q \end{array}$$

Take $W = P \cup T$. To show this works, the key point is to show the image of

$$CH^2(T_{0\mathbf{C}}) \to CH^2(S_{0\mathbf{C}})$$

is "small". One can do this by using cycle classes in $H^2(\ \ ,\Omega^2_{\bullet/\overline{\mathbf{C}}})$:

$$\begin{array}{ccc}
CH^2(T_{0\mathbf{C}}) & \longrightarrow & CH^2(S_{0\mathbf{C}}) \\
\downarrow & & \downarrow \\
H^2(T_{0\mathbf{C}}, \Omega^2_{T_{0\mathbf{C}}/\overline{\mathbf{Q}}}) & \longrightarrow & H^2(S_{0\mathbf{C}}, \Omega^2_{S_{0\mathbf{C}}/\overline{\mathbf{Q}}}) \\
\downarrow & & \downarrow \\
(0) = H^2(T_{0\mathbf{C}}, \mathcal{O}_{T_{0\mathbf{C}}}) \otimes_{\mathbf{C}} \Omega^2_{\mathbf{C}/\overline{\mathbf{Q}}} & \longrightarrow & H^2(S_{0\mathbf{C}}, \mathcal{O}_{S_{0\mathbf{C}}}) \otimes \Omega^2_{\mathbf{C}/\overline{\mathbf{Q}}}.
\end{array}$$

Mumford's differential techniques for showing $CH^2(S_{0\mathbf{C}})$ is large can be reinterpreted to prove that the image of $d(x) - h^2$ in $H^2(S_0, \mathcal{O}_{S_0}) \otimes \Omega^2_{\mathbf{C}/\overline{\mathbf{Q}}}$ is non-zero for x $\overline{\mathbf{Q}}$-generic.

Best,

Spencer Bloch

Appendix B : An example by C. Schoen

An example is given of the following phenomenon: A smooth projective surface V over a field k satisfying:

i)   rank $(\mathrm{Ker}: CH_0(V)_{\deg 0} \longrightarrow \mathrm{Alb}_V(k)) = \infty$.

ii) V cannot be obtained by base changing a variety V'/k' for a field $k' \subset k$ with trans. deg. $(k/k') > 0$.

Begin with the elliptic curve $E/\overline{\mathbb{Q}}$ with equation $zy^2 = x^3 + \frac{1}{4}z^3$ and origin $e = (0:1:0)$. The group of cube roots of unity, μ_3 , acts on E via multiplication on the x–coordinate. Thus $(\mu_3)^3$ acts on E^3 . The largest subgroup $H \lhd (\mu_3)^3$ which acts trivially on $H^0(E^3,\Omega^3) \cong H^0(E,\Omega^1)^{\otimes 3}$ contains the diagonal $\mu_3 \cong \Delta \lhd \mu_3^3$ as an index 3 subgroup. In fact, if $M \lhd H$ is the largest subgroup which operates trivially on the first factor, then $H = M \times \Delta$.

Projection on the first factor $\mathrm{pr}_1 : E^3 \longrightarrow E$ induces a commutative diagram of morphisms

$$
\begin{array}{ccccc}
E^3 & \longrightarrow & E^3/\Delta & \longrightarrow & E^3/H \\
{\scriptstyle \mathrm{pr}_1}\downarrow & & {\scriptstyle \overline{\mathrm{pr}}_1}\downarrow & & {\scriptstyle \overline{\overline{\mathrm{pr}}}_1}\downarrow \\
E & \xrightarrow{\ f\ } & E/\mu_3 \cong \mathbb{P}^1 & = & E/\mu_3 \cong \mathbb{P}^1 \\
\end{array} \quad ,
$$

in which f is the canonical quotient map. Let $k = \overline{\mathbb{Q}}(\mathbb{P}^1) \cong \overline{\mathbb{Q}}(t)$. Write F_k (respectively T_k) for the generic fiber of $\overline{\mathrm{pr}}_1$ (respectively $\overline{\overline{\mathrm{pr}}}_1$) . Then F_k is an abelian surface, and $T_k \cong F_k/(H/\Delta)$ is a singular K3 surface whose singularities are resolved by blowing up $(T_k)_{\mathrm{sing}}$. Let V_k denote the resulting non–singular K3 surface. Thus $\mathrm{Alb}_V = 0$.

To check that (ii) is satisfied it suffices to show that V_k is not the base change of a variety $V'/\overline{\mathbb{Q}}$. If such V' were to exist, $\mathrm{Gal}(\overline{k}/k)$ would act trivially on $H^2(V_{\overline{k}},\mathbb{Q}_\ell) \cong H^2(V',\mathbb{Q}_\ell)$. We claim however that $\mathrm{Gal}(\overline{k}/k)$ does not even act trivially on the quotient $H^2(F_{\overline{k}},\mathbb{Q}_\ell)^{H/\Delta}$. Since the base change of F_k by the field extension $\overline{\mathbb{Q}}(E)/k$ yields the constant abelian surface $E \times E$ over $\overline{\mathbb{Q}}(E)$, the Galois action factors through $\mathrm{Gal}(\overline{\mathbb{Q}}(E)/k)$. The natural action of this group on $F_k \times_k \overline{\mathbb{Q}}(E)$ is induced by an isomorphism $\mathrm{Gal}(\overline{\mathbb{Q}}(E)/k) \cong \Delta \subseteq \mathrm{Aut}(E^3_{\overline{\mathbb{Q}}})$. Since Δ acts non–trivially on $[\Lambda^2 H^2(e \times E \times E_{\overline{\mathbb{Q}}},\mathbb{Q}_\ell)]^M$, $\mathrm{Gal}(\overline{\mathbb{Q}}(E)/k)$ acts non–trivially on $H^2(F_{\overline{k}},\mathbb{Q}_\ell)^{H/\Delta}$.

For any variety $W/\overline{\mathbb{Q}}$ write $B_1(W)$ for the group of 1–cycles modulo algebraic equivalence. Let $\gamma : U \longrightarrow E^3/H$ be a resolution of singularities such that the generic fiber of $p := \overline{\overline{\mathrm{pr}}}_1 \circ \gamma$ is isomorphic to V_k. It is possible to deduce (i) if one knows that rank $B_1(U_{\overline{\mathbb{Q}}}) = \infty$. In fact,

$$CH_0(V_k) = \lim_{D} CH_1(U_{\overline{\mathbb{Q}}} - p^{-1}(D))$$

as D ranges over reduced effective divisors on $\mathbb{P}^1_{\overline{\mathbb{Q}}}$. Hence, by the commutative diagram with exact row

$$\begin{array}{ccccccc} & & & & \lim\limits_{D}\ CH_1(U_{\overline{\mathbb{Q}}} - p^{-1}(D)) & & \\ & & & & \big\downarrow & & \\ \lim\limits_{D} B_1(p^{-1}(D)) & \xrightarrow{\ c\ } & B_1(U_{\overline{\mathbb{Q}}}) & \longrightarrow & \lim\limits_{D}\ B_1(U_{\overline{\mathbb{Q}}} - p^{-1}(D)) & \longrightarrow & 0 \end{array}$$

one need only to show that the image of c has finite rank. This is true because an open subset of U is dominated by a constant family of surfaces. More precisely, there are birational morphisms of non–singular projective $\overline{\mathbb{Q}}$–varieties $\xi : Z \longrightarrow E^2$ and $\sigma : Y \longrightarrow E \times Z$ where

(α) the exceptional locus of σ maps to a finite subset of E,
(β) there is a commutative diagram of morphisms

$$\begin{array}{ccc} Y & \xrightarrow{\ \tilde{f}\ } & U \\ {\scriptstyle (\mathrm{id} \times \xi) \circ \sigma} \big\downarrow & & \big\downarrow {\scriptstyle \gamma} \\ E^3 & \xrightarrow{\ \mathrm{can.}\ } & E^3/H . \end{array}$$

Let $\overset{\circ}{\mathbb{P}} \subset \mathbb{P}^1$ be a non–empty Zariski open subset over which p and $q := f \circ \mathrm{pr}_1 \circ (\mathrm{id} \times \xi) \circ \sigma = p \circ \tilde{f}$ are smooth. Denote the base change to $\overset{\circ}{\mathbb{P}}$ of an object over $\mathbb{P}^1$ by adding $\circ$ to the notation. Thus $\overset{\circ}{Y} \cong \overset{\circ}{E} \times Z$, and restriction $B_1(U) \longrightarrow B_1(\overset{\circ}{U})$ is surjective with finitely generated kernel. The vertical maps in the following commutative diagram are surjective

$$\begin{array}{ccc}
\lim\limits_{D\subset\overset{\circ}{\mathbb{P}}} B_1(\overset{\circ}{q}{}^{-1}(D))\otimes\mathbb{Q} & \xrightarrow{\ c'\ } & B_1(\overset{\circ}{E}\times Z)\otimes\mathbb{Q} \\
\downarrow & & \downarrow (\Upsilon)^\circ \\
\lim\limits_{D\subset\overset{\circ}{\mathbb{P}}} B_1(\overset{\circ}{p}{}^{-1}(D))\otimes\mathbb{Q} & \xrightarrow{\ \overset{\circ}{c}\ } & B_1(\overset{\circ}{U})\otimes\mathbb{Q}
\end{array}$$

(note that $(\Upsilon)^\circ$ is proper and surjective). Because $\overset{\circ}{E}\times Z/\overset{\circ}{E}$ is a constant family, c' has finite dimensional image: by the definition of algebraic equivalence, corresponding cycles in different fibers of $\overset{\circ}{E}\times Z/\overset{\circ}{E}$ have the same image in $B_1(\overset{\circ}{E}\times Z)$, and the Néron–Severi group of Z is finitely generated. Thus c has finite rank image as desired.

Finally, the fact that $\operatorname{rank} B_1(U_{\overline{\mathbb{Q}}}) = \infty$ may be deduced from a similar statement for the elliptic modular 3–fold $\tilde{W}(3)_{\overline{\mathbb{Q}}}$ [1, p. 778]. The correspondence between $\tilde{W}(3)_{\overline{\mathbb{Q}}}$ and $E^3_{\overline{\mathbb{Q}}}$ constructed in [2, § 1] (where $\tilde{W}(3)$ is called $\tilde{W}$) gives rise to a correspondence $Q\in CH_3(\tilde{W}(3)\times U_{\overline{\mathbb{Q}}})$ such that

$$Q^*: H^0(U_{\overline{\mathbb{Q}}},\Omega^3)\longrightarrow H^0(\tilde{W}(3)_{\overline{\mathbb{Q}}},\Omega^3)$$

is an isomorphism. Since these vector spaces are not zero (in fact they are one dimensional) [1, Thm. 4.7] implies that $\operatorname{rank} B_1(U_{\overline{\mathbb{Q}}}) = \infty$.

[1] C. Schoen: Complex multiplication cycles on elliptic modular threefolds, Duke Math. J. 53 (1986), 771–794.

[2] C. Schoen: Zero cycles modulo rational equivalence for some varieties over fields of transcendence degree one, Proc. Symp. Pure Math. 46 (Algebraic Geometry, Bowdoin 1985), Part 2, A.M.S., p. 463–473.

Appendix C: Complements and problems

C1. I was asked to add some clarifying words on the weight filtration on ℓ-adic cohomology. Let U be a smooth variety over a finitely generated field k of characteristic zero. It was not mentioned but is true that the filtration on $H^n_{ét}(\overline{U},\mathbb{Q}_\ell)$ constructed in 3.14 is a weight filtration in the sense of 6.8. Namely, with the notation of § 3, $Gr^W_m H^n_{ét}(\overline{U},\mathbb{Q}_\ell)$ is a subquotient of $H^{2n-m}_{ét}(\overline{Y^{(m-n)}},\mathbb{Q}_\ell(n-m))$ as a G_k-module by 3.20, and this is pure of weight m, since $Y^{(n-m)}$ is smooth and proper.

That $H^i(\overline{X},\mathbb{Q}_\ell)$ is pure of weight i, for X smooth and proper over k, follows as in 7.12: Choose a smooth and proper extension $f: \mathcal{X} \longrightarrow S$ over a model S of k (an integral scheme of finite type over $\mathbb{Z}[1/\ell]$, with function field k). Smooth and proper base change gives a $Gal(\overline{\kappa(y)}/\kappa(y))$-isomorphism

$$H^i(\overline{X},\mathbb{Q}_\ell) \cong H^i(\mathcal{X}\times_S \overline{\kappa(y)},\mathbb{Q}_\ell)$$

for every $y \in S$. If $x \in S$ is closed, then Deligne's proof of the Weil conjectures for the smooth and proper variety $\mathcal{X}\times_S \kappa(x)$ over the finite field $\kappa(x)$ ([D 8] (1.6), as amplified by [D 9] (3.3.9)) shows that the eigenvalues of Fr_x on the above cohomology group are pure of weight i.

By 6.8.2 there are weight filtrations with the correct weights (i.e., those of 6.5) on $H^i_{ét,Z}(\overline{X},\mathbb{Q}_\ell)$ and $H_a(\overline{X},\mathbb{Q}_\ell)$ for arbitrary varieties X and closed subvarieties $Z \subset X$ over any finitely generated field k of characteristic $\neq \ell$. These weight filtrations are unique and functorial by 6.8.1. In particular, they coincide with the one in 3.14 in the case above. This also follows from the construction in 6.11.

C2. It is a problem what realizations one has to take to get the good category of mixed motives. Probably it is not enough to consider smooth varieties as in 4.1. Perhaps one also needs singular varieties (see C3 below) or other "geometric" realizations like those constructed by Deligne in [D 11].

The construction of 6.11 attaches "cohomological" and "homological" realizations to arbitrary varieties, morphisms of varieties and, most generally (as noted in 6.11), to simplicial varieties. The technique of § 4 always gives Tannakian categories and associated "Galois" groups with the properties stated in 4.4 and 4.7, provided one takes a Tannakian subcategory of $\underline{MR}_k$ generated by a class of mixed realizations with the following two properties:

a) It contains all $H^n(X)$ for X smooth and projective,
b) The pure quotients are in $\underline{M}_k$ (and hence polarizable).

This is still the case for the realizations attached to simplicial varieties.

Recall from 11.3 that for any field k one would like to define a $\mathbb{Q}$–linear tensor category with weights $\mathcal{MM} = \mathcal{MM}_k$ such that

i) the pure objects are semi–simple, and the subcategory of semi–simple objects is equivalent to Grothendick's category $\mathcal{M} = \mathcal{M}_k$ of (pure) motives with respect to numerical equivalence over k,

ii) there is a twisted Poincaré duality theory with weights $(H^i_Z(X,j), H_a(X,b))$ with values in $\mathcal{MM}$, and

iii) there is a spectral sequence ($\underline{1}$ = identity object)

$$E_2^{p,q} = \mathrm{Ext}^p_{\mathcal{MM}}(\underline{1}, H^q(X,j)) \Rightarrow H^{p+q}_{\mathcal{M}}(X,\mathbb{Q}(j))$$

converging to the motivic cohomology defined via K–theory and degenerating for a smooth and projective variety.

We can add here that iv) for a field k of arithmetical dimension d the cohomological dimension of $\mathcal{MM}$ should be d, i.e., one should have $\mathrm{Ext}^p_{\mathcal{MM}} = 0$ for $p > d$. For a finite field this would mean that $\mathcal{MM}$ coincides with Grothendieck's category of motives. For a global field this would imply the short exact sequence

$$0 \longrightarrow \mathrm{Ext}^1_{\mathcal{MM}}(\underline{1}, H^{i-1}(X,j)) \longrightarrow H^i_{\mathcal{M}}(X,\mathbb{Q}(j)) \longrightarrow \mathrm{Hom}_{\mathcal{MM}}(\underline{1}, H^i(X,j)) \longrightarrow 0$$

mentioned in 11.4 c). In contrast to this, for a smooth projective surface X over $\mathbb{C}$ one expects a non–trivial group

$$\mathrm{Ext}^2_{\mathcal{MM}_{\mathbb{C}}}(\underline{1}, H^2(X,2)) = T(X)$$

for $p_g(X) \neq 0$. To deduce this formula from the spectral sequence iii) note that

$$\mathrm{Ext}^4_{\mathcal{MM}}(\underline{1}, H^0(X,2)) = \mathrm{Ext}^4_{\mathcal{MM}}(\underline{1}, \underline{1}(2)) \subseteq H^4_{\mathcal{M}}(\mathrm{Spec}\,\mathbb{C}, \mathbb{Q}(2)) = CH^2(\mathrm{Spec}\,\mathbb{C})_{\mathbb{Q}} = 0$$

and

$$\mathrm{Ext}^3_{\mathcal{MM}}(\underline{1}, H^1(X,2)) \subseteq \mathrm{Ext}^3_{\mathcal{MM}}(\underline{1}, H^1(C,2)) \subseteq CH^2(C)_{\mathbb{Q}} = 0,$$

using a curve C with $H^1(X) \hookrightarrow H^1(C)$.

For a field k of characteristic zero, the category $\underline{M}_k$ constructed by absolute Hodge cycles coincides with $\mathcal{M}_k$ if (and only if) every absolute Hodge cycle is algebraic (and this would follow from either the Hodge or the Tate conjecture, cf. 5.4). Therefore the category $\underline{MM}_k$ of 4.1 is expected to satisfy i), but I don't know if it satisfies ii). Certainly it satisfies ii) (and still i)) after suitable enlargement, e.g., by adding all mixed realizations in the image of the twisted Poincaré duality theory of 6.11.1, but it seems hard to guess which enlargement should satisfy iii) and iv).

Of course it would be desirable to have a purely algebraic description of $\mathcal{MM}_k$, even a conjectural one, and perhaps this is not out of reach: the idea would be to use Beilinson's filtration on the Chow groups (cf. § 11) to construct $\mathcal{MM}_k$ as the heart of a certain triangulated category. In any case one conjectures that the realization functor H to $\underline{MR}_k$ is fully faithful and identifies $\mathcal{MM}_k$ with a Tannakian subcategory of $\underline{MR}_k$, which makes the construction of § 4 reasonable. For a characterization of the image of H, i.e., of the motivic realizations, it will be useful to consider further realization functors (e.g., crystalline ones as in Grothendieck's original approach to motives) and further comparison isomorphisms (e.g., those conjectured in [Fo] and partly proved in [FM] and [Fa]). While it seems a complete mystery how to characterize the pure motivic realizations it may be easier to predict those extensions of given pure motives which are motivic, cf. [Bei 1] and [BK].

C3. The spectral sequence iii) of the previous section arises the question for the relation between motivic cohomology and motivic extensions, i.e., extensions in the category $\underline{MM}_k$ of 4.1 (or similar ones). First consider zero–extensions, i.e., motivic morphisms. The Chern characters give maps

$$ch_{i,j} : H^i_{\mathcal{M}}(X,\mathbb{Q}(j)) \longrightarrow \Gamma H^i_{AH}(X,j) = \mathrm{Hom}_{\underline{MR}_k}(\underline{1},H^i_{AH}(X,j)) ,$$

and by definition the target group is $\mathrm{Hom}_{\underline{MM}_k}(\underline{1},H^i_{AH}(X,j))$ if $H^i_{AH}(X,j)$ is in $\underline{MM}_k$ (e.g., if X is smooth). Hence $ch_{i,j}$ can be regarded as the first edge morphism in iii), and the degeneration of iii) for smooth and projective X is related to the surjectivity of $ch_{i,j}$ for $i = 2j$ (the only case where $\Gamma H^i_{AH}(X,j) \neq 0$), which was discussed in 5.4.

Now let $H^i_{\mathcal{M}}(X,\mathbb{Q}(j))_0 = \ker ch_{i,j}$. The constructions of 6.11 give maps

$$ch'_{i,j} : H^i_{\mathcal{M}}(X,\mathbb{Q}(j))_0 \longrightarrow \mathrm{Ext}^1_{\underline{MR}_k}(\underline{1},H^{i-1}_{AH}(X,j)) .$$

I have not shown that these have image in

$$\mathrm{Ext}^1_{\underline{MM}_k}(\underline{1},H^{i-1}_{AH}(X,j)) \subseteq \mathrm{Ext}^1_{\underline{MR}_k}(\underline{1},H^{i-1}_{AH}(X,j)) ,$$

at least not in general. Only for smooth X and $i = 2j$ the results in § 9 show that one gets motivic extensions: for the considered Poincaré duality theory the object E in 9.1.1 is in $\underline{MM}_k$ as a subobject of $H^{2j-1}_{AH}(U,j)$ (cf. 4.2). It is possible to extend this somewhat to the case $i > j$.

For example, let X be smooth of dimension d, and let z be an element of $H^1(X, \mathcal{K}_2)$. It is represented by a finite family (f_i), with $f_i \neq 0$ in the function field of an irreducible divisor $Y_i \subset X$ and $\sum_i \operatorname{div}(f_i) = 0$ on X (cf. 6.12.4 e)). If $Y = \bigcup_i Y_i$, then (f_i) defines an element in $E^2_{d-1,d+2}(Y)$, hence an element in $H^{ét}_{2d-3}(Y, \mathbb{Z}_\ell(d-2)) \cong H^3_{ét,Y}(X, \mathbb{Z}_\ell(2))$ (cf. 8.13). If it is mapped to zero in $H^3_{ét}(\overline{X}, \mathbb{Z}_\ell(2))$, one obtains a motivic extension of $\mathbb{Z}_\ell$ by $H^2_{ét}(\overline{X}, \mathbb{Z}_\ell(2))/N^Y$ via pull-back from the exact sequence

$$0 \longrightarrow H^2_{ét}(\overline{X}, \mathbb{Z}_\ell(2))/N^Y \longrightarrow H^2_{ét}(\overline{X-Y}, \mathbb{Z}_\ell(2)) \longrightarrow H^3_{ét,Y}(X, \mathbb{Z}_\ell(2)),$$

where N^Y is the image of $H^2_{ét,Y}(X, \mathbb{Z}_\ell(2))$ as in 9.16. This construction amounts to the one communicated to me by A. Scholl in a letter from September 1985 (cf. the introduction). One can show that the element in $H^3_{ét}(\overline{X}, \mathbb{Q}_\ell(2))$ associated to z by the above prescription is $ch_{3,2}(z)$ via the identfication $H^1(X, \mathcal{K}_2) \otimes \mathbb{Q} = H^3_{\mathcal{M}}(X, \mathbb{Q}(2))$ of 6.12.4 e) and that, for $z \in H^3_{\mathcal{M}}(X, \mathbb{Q}(2))_0$, $ch'_{3,2}(z) \in \operatorname{Ext}^1_{G_k}(\mathbb{Q}_\ell, H^2(\overline{X}, \mathbb{Q}_\ell(2)))$ induces the above extension via push-out to $H^2(\overline{X}, \mathbb{Q}_\ell(2))/N^Y$ (e.g., by using 9.5).

More generally, for $i > j$ every $z \in H^i_{\mathcal{M}}(X, \mathbb{Q}(j))$ comes from an element $z' \in H^i_{\mathcal{M},Y}(X, \mathbb{Q}(j))$ for some $Y \subset X$ of codimension $\geq i-j$ (cf. 5.23). If $ch_{i,j}(z) = 0$ in $\Gamma H^i_{AH}(X,j)$, then the image of $ch'_{i,j}(z)$ in $\operatorname{Ext}^1_{\underline{MR}_k}(\underline{1}, H^{i-1}_{AH}(X,j)/N^Y)$ is the class of the pull-back extension

$$\begin{array}{ccccccccc}
0 \longrightarrow H^{i-1}_{AH}(X,j)/N^Y & \longrightarrow & H^{i-1}(X-Y,j) & \longrightarrow & H^i_{AH,Y}(X,j) & \longrightarrow & H^i_{AH}(X,j) \\
\| & & \cup\!| & & \uparrow z' & & \\
0 \longrightarrow H^{i-1}_{AH}(X,j)/N^Y & \longrightarrow & E & \longrightarrow & \underline{1} & \longrightarrow & 0\ ,
\end{array}$$

(where $N^Y = \operatorname{Im}(H^{i-1}_{AH,z}(X,j) \longrightarrow H^{i-1}_{AH}(X,j))$ as in 9.16) and hence motivic. The class obviously vanishes if z is in

$$\tilde{N}^Y H^i_{\mathcal{M}}(X, \mathbb{Q}(j)) = \operatorname{Im}(H^i_{\mathcal{M},Y}(X, \mathbb{Q}(j))_{00} \longrightarrow H^i_{\mathcal{M}}(X, \mathbb{Q}(j))),$$

where

$$H^i_{\mathcal{M},Y}(X,\mathbb{Q}(j))_{00} = \mathrm{Ker}(H^i_{\mathcal{M},Y}(X,\mathbb{Q}(j)) \longrightarrow \Gamma H^i_{AH,z}(X,j)) .$$

We thus obtain maps (for $\nu \geq 0$)

$$H^i_{\mathcal{M}}(X,\mathbb{Q}(j))_0/\tilde{N}^\nu \longrightarrow \mathrm{Ext}^1_{\underline{MM}_k}(\underline{1},H^{i-1}_{AH}(X,j)/N^\nu) ,$$

where $N^\cdot$ is the coniveau filtration and $\tilde{N}^\nu = \cup \tilde{N}^Y$, with Y running through all closed subvarieties of codimension ν in X, cf. 9.16 (where $\tilde{N}^Y$ and $\tilde{N}^\nu$ were also denoted N^Y and N^ν, with a risk of confusion). If k is a number field and X is smooth and proper, then one expects injectivity of

$$H^i_{\mathcal{M}}(X,\mathbb{Q}(j))_0/\tilde{N}^Y \longrightarrow \mathrm{Ext}^1_{\underline{MM}_k}(\underline{1},H^{i-1}_{AH}(X,j)/N^Y)$$

and the above maps by similar arguments as in 9.16. By using a splitting $H^{i-1}_{AH}(X,j) = N^Y \oplus H^{i-1}_{AH}(X,j)/N^Y$ one then has to study if

$$ch'_{i,j} : H^i_{\mathcal{M},Y}(X,\mathbb{Q}(j)) \longrightarrow \mathrm{Ext}^1_{\underline{MR}_k}(\underline{1},H^{i-1}_{AH,Y}(X,j))$$

has image in the motivic extensions, but here the above method will not apply in general: For example, if $(i,j) = (4,3)$ and Y is a smooth irreducible division, then by Poincaré duality isomorphism we study

$$H^2_{\mathcal{M}}(Y,\mathbb{Q}(2)) \longrightarrow \mathrm{Ext}^1_{\underline{MR}_k}(\underline{1},H^1_{AH}(Y,2)) ,$$

i.e., a situation with $i \leq j$.

Concerning the case $i = j = 1$ one has $H^1_{\mathcal{M}}(X,\mathbb{Q}(1))_0 \xleftarrow[\sim]{p^*} H^1_{\mathcal{M}}(\mathrm{Spec}\, k,\mathbb{Q}(1))$ for a smooth geometrically connected variety $X \xrightarrow{p} \mathrm{Spec}\, k$, and one may use 1–motives to construct geometric extensions associated to elements in $H^1_{\mathcal{M}}(\mathrm{Spec}\, k,\mathbb{Q}(1)) = k^\times \otimes \mathbb{Q}$: One can show that for $x \in k^\times$ the 1–motive ([D5] 10.1)

$$M: \quad \begin{array}{c} \mathbb{Z} \longrightarrow \mathbb{G}_m \\ 1 \longmapsto x \in \mathbb{G}_m(k) = k^\times \end{array}$$

gives rise to an extension of realizations

$$0 \longrightarrow \underline{1}(1) \longrightarrow T(M)_{\mathbb{Q}} \longrightarrow \underline{1} \longrightarrow 0 \, ,$$

whose class in $\mathrm{Ext}^1_{\underline{MR}_k}(\underline{1},\underline{1}(1))$ is $\mathrm{ch}'_{1,1}(x)$. $T(M)_{\mathbb{Q}}$ is also obtained as relative cohomology of smooth varieties or as cohomology of a singular, non—proper variety, by the isomorphism $T(M)_{\mathbb{Q}} \cong H^1(\mathbb{G}_m(x),1)$ and the commutative exact diagram

$$
\begin{array}{ccccccccc}
\underline{1}(1) & \xrightarrow{\text{diag.}} & \underline{1}(1)^2 & \longrightarrow & H^1(\mathbb{G}_m \mathrm{mod}\{1,x\};1) & \longrightarrow & H^1(\mathbb{G}_m,1) & \longrightarrow & 0 \\
\downarrow & & \downarrow{\scriptstyle (a,b)} & & \downarrow \int & & \downarrow \int & & \\
& & \downarrow{\scriptstyle a-b} & & & & & & \\
0 & \longrightarrow & \underline{1}(1) & \longrightarrow & H^1(\mathbb{G}_m(x),1) & \longrightarrow & \underline{1} & \longrightarrow & 0 \, ,
\end{array}
$$

where $\mathbb{G}_m(x)$ is $\mathbb{G}_m = \mathrm{Spec}(k[t,t^{-1}])$ with 1 and x glued together. The isomorphism follows with the methods of [D 5] 10.3. I don't know if $T(M)_{\mathbb{Q}}$ is an object of $\underline{MM}_k$ as defined in 4.1.

Concerning $i < j$, I do not even know if the extensions of $\underline{1}$ by $\underline{1}(j)$ attached to elements in $H^1_{\mathcal{M}}(\mathrm{Spec}\ k,\mathbb{Q}(j))$ by $\mathrm{ch}'_{i,j}$ appear in some kind of cohomological realization for $j > 1$. In [D 11] Deligne has shown that certain canonical extensions appear in the realizations attached to $\pi_1(\mathbb{P}^1_{\overline{\mathbb{Q}}}\backslash\{0,1,\infty\})$, but it is unknown, if they come from some elements in K—theory. It may be possible to use Bloch's description of $\mathrm{ch}_{i,j}$ via cycle maps on higher Chow groups [Bl 6] and a construction analogous to the one in § 9 to produce geometric extensions from elements in $H^i_{\mathcal{M}}(X,\mathbb{Q}(j))$ for arbitrary i and j.

Of course, it is as important to study the relation between motivic cohomology and extensions in the single realizations. For example, for a smooth, projective variety X over a number field k one expects that the map

$$\mathrm{ch}'_{i,j} \otimes \mathbb{Q}_\ell : H^i_{\mathcal{M}}(X,\mathbb{Q}(j))_0 \otimes \mathbb{Q}_\ell \longrightarrow \mathrm{Ext}^1_{G_k}(\mathbb{Q}_\ell,H^{i-1}_{\mathrm{\acute{e}t}}(\overline{X},\mathbb{Q}_\ell(j))) = H^1(G_k,H^{i-1}_{\mathrm{\acute{e}t}}(\overline{X},\mathbb{Q}_\ell(j))$$

is injective and that its image can be described by explicit local conditions (see [BK] (5.3), and [J3] § 6 as well as a forthcoming paper; this extends the conjectures in § 13 to all $i,j \in \mathbb{Z}$). This in turn would imply that the map

$$\mathrm{Ext}^1_{\mathcal{M}\mathcal{M}_k}(\underline{1},H^i(X,j)) \longrightarrow \mathrm{Ext}^1_{G_k,\mathrm{mot}}(\mathbb{Q}_\ell,H^i_{\mathrm{\acute{e}t}}(\overline{X},\mathbb{Q}_\ell(j)))$$

obtained by "passing to the ℓ-adic realizations" is an isomorphism, where the "motivic G_k-extensions" forming the target group are characterized by local properties related to Fontaine's theory of p-adic representations.

C4. It may be useful to recall how one can calculate the Yoneda-Ext-groups in a neutral Tannakian category in terms of group cohomology of the associated "Galois" group: Let G be a linear algebraic group over a field of characteristic zero, let U be its unipotent radical, and let Rep G be the category of finite dimensional algebraic representations of G (these are the rational modules of [Ho 1]). Then for objects V, W in Rep G one has

$$\mathrm{Ext}^i_{\underline{\mathrm{Rep}}\,G}(V,W) = H^i(G,\underline{\mathrm{Hom}}(V,W)),$$

where $H^i(G,-)$ is the cohomology theory defined in [Ho 1], and isomorphisms

$$H^i(G,V) \xrightarrow{\sim} H^i(U,V)^{G/U},$$

$$H^i(U,V) \xrightarrow{\sim} H^i(u,V),$$

where $u = \mathrm{Lie}\, U$ is the Lie algebra of U and $H^i(u,-)$ denotes Lie algebra cohomology ([Ho 2]).

All this extends to pro-algebraic groups and continuous representations by passing to the limit. Putting this together we obtain the isomorphisms

$$\mathrm{Ext}^\nu_{\underline{MM}_k}(\underline{1},M) \xrightarrow{\sim} H^\nu(\mathrm{Lie}\; U(\sigma),M_\sigma)^{MG(\sigma)},$$

where M_σ is a $MG(\sigma)$-module by definition.

Let me also mention that one may use injectives to calculate the Yoneda-Ext-groups, even if $\underline{MM}_k$ does not have enough injectives. But for any neutral Tannakian category Rep G the ind-category has enough injectives and these may be used to calculate $H^i(G,-)$. In fact, these are the rationally injective modules of [Ho 1].

C5. Let $\underline{MM}_k$ be defined as in 4.1 or as a suitable enlargement. While it seems out of reach to describe this category completely, i.e., to determine the whole group $MG(\sigma)$ for one $\sigma : k \hookrightarrow \mathbb{C}$, it may be possible to describe certain subcategories, i.e., to calculate certain quotients of $MG(\sigma)$. By Deligne's result that every Hodge cycle on an abelian variety A is absolute Hodge ([DMOS] I 2.11), one can in principle determine the Tannakian subcategory of $\underline{M}_k$ generated by $H^1(A)$ in terms of the Mumford-Tate group of A. In any case, one has a nice description of the category $\underline{PCM}_{\mathbb{Q}}$ of motives of potential CM-type over $\mathbb{Q}$ (generated by

Artin motives and abelian varieties of potential CM–type), in terms of Hecke characters and the Taniyama group ([DMOS] IV).

In the mixed case, Brylinski has proved an analogue of Deligne's theorem for a 1–motive M ([Br] 2.2.5). However, it is not clear to me whether his theorem is sufficient for the determination of the quotient of $MG(\sigma)$ corresponding to the Tannakian category generated by $T(M)_{\mathbb{Q}}$ and $\underline{1}(1)$. Brylinski only considers Hodge and absolute Hodge cycles in tensor products

$$T(M)_{\mathbb{Q}}^{\otimes p} \otimes T(M)_{\mathbb{Q}}^{\vee \otimes q} \otimes \underline{1}(r)$$

whereas for non–reductive groups one a priori has to consider cycles in subquotients of such spaces, too ([DMOS] I 3.2 (a)).

Another object of interest is the category $\mathcal{MT} = \mathcal{MT}_k$ of mixed Tate motives over k : these are mixed motives whose pure quotients are sums of pure Tate motives $\underline{1}(r)$, $r \in \mathbb{Z}$. There are now several proposals for an algebraic description of such a category ([BMS], [BGSV]). It may be interesting to compare them with other candidates $\underline{MT}_k$ constructed in $\underline{MR}_k$ by the techniques of § 4. I don't know how close the category $\underline{L}_k$ generated by realizations of linear varieties (see § 14) comes to such a category $\underline{MT}_k$.

Suppose $\mathcal{MM}$ and $\mathcal{MT}$ exist. What are the properties of the maps

$$\mathrm{Ext}^i_{\mathcal{MT}}(\underline{1},\underline{1}(j)) \longrightarrow \mathrm{Ext}^i_{\mathcal{MM}}(\underline{1},\underline{1}(j))$$

for $j > 1$? Will the expected maps

$$H^i_{\mathcal{M}}(\mathrm{Spec}\, k,\mathbb{Q}(j)) \longrightarrow \mathrm{Ext}^i_{\mathcal{MM}}(\underline{1},\underline{1}(j))$$

factorize through $\mathrm{Ext}^i_{\mathcal{MT}}$ for $i > 1$? In particular, can one associate successive extensions of Tate realizations (e.g., ℓ–adic ones) to elements in K–groups of k ?

Bibliography

[BFM] P. Baum, W. Fulton, R. MacPherson: Riemann-Roch for singular varieties, Publ. Math. I.H.E.S. 45 (1975), 101-145.

[Bei 1] A. Beilinson: Higher regulators and values of L-functions, J. Soviet Math. 30 (1985), 2036-2070.

[Bei 2] A. Beilinson: Notes on absolute Hodge cohomology; Applications of Algebraic K-Theory to Algebraic Geometry and Number Theory, Contemporary Mathematics 55 I, A.M.S., Providence 1986, p. 35-68.

[Bei 3] A. Beilinson: Letter to Soulé, 1/11/82.

[Bei 4] A. Beilinson: Height pairings between algebraic cycles, Springer Lecture Notes 1289 (1987), p. 1-26.

[BGSV] A. Beilinson, A. Goncharov, V. Schechtman, A. Varchenko: Aomoto dilogarithms, mixed Hodge structures and motivic cohomology of pairs of triangles on the plane, preprint 1989.

[Ber] P. Berthelot, Cohomologie Cristalline des Schémas de Caractéristique p>0, Springer Lecture Notes 407 (1974).

[BMS] A. Beilinson, R. MacPherson, V. Schechtman: Notes on motivic cohomology, Duke Math. J. 54 (1987), 679-710.

[Bl 1] S. Bloch: Lectures on Algebraic Cycles, Duke University Math. Series IV, Durham 1980.

[Bl 2] S. Bloch: Some elementary theorems about cycles on abelian varieties, Invent. Math. 37 (1976), 215-228.

[Bl 3] S. Bloch: Algebraic cycles and values of L-functions, J. reine angew. Math. 350 (1984), 94-108.

[Bl 4] S. Bloch: Algebraic cycles and values of L-functions II, Duke Math. J. 52 (1985), 379-397.

[Bl 5] S. Bloch: Algebraic cycles and higher K-theory, Advances in Math. 61 (1986), 267-304.

[Bl 6] S. Bloch: Algebraic cycles and the Beilinson conjectures; The Lefschetz Centennial Conference, Contemporary Mathematics 58 I, A.M.S., Providence 1986, p. 65-79.

[BK] S. Bloch, K. Kato: L-functions and Tamagawa numbers of motives, preprint 1988.

[BO] S. Bloch, A. Ogus: Gersten's conjecture and the homology of schemes, Ann. Sci. ENS (4) 7 (1979), 181-202.

[BS] S. Bloch, V. Srinivas: Remarks on correspondences and algebraic cycles, Amer. J. Math. 105 (1983), 1235-1253.

[Br] J.-L. Brylinski: "1-Motifs" et formes automorphes (Théorie arithmétique des domaines de Siegel); Journées Automorphes, Publ. Math. de l'Univ. Paris VII, 15 (1983), 43-106.

[Ca] J. Carlson: Extensions of mixed Hodge structures; Journées de geométrie algébrique d'Angers 1979, Sijthoff & Noordhoff 1980, p. 107-127.

[C] J.-L. Colliot-Thélène: Hilbert's theorem 90 for K_2, with application to the Chow groups of rational surfaces, Invent. Math. 71 (1983), 1-20.

[CR] J.-L. Colliot-Thélène, W. Raskind: K_2-cohomology and the second Chow group, Math. Ann. 270 (1985), 165-199.

[CSS] J.-L. Colliot-Thélène, J.-J. Sansuc, C. Soulé: Torsion dans le groupe de Chow de codimension deux, Duke Math. J. 50 (1983), 763-801.

[D 1] P. Deligne: Formes modulaires et représentations ℓ-adiques; Sem. Bourbaki 1968/69 exp. 355, Springer Lecture Notes 179 (1971), p. 139-172.

[D 2] P. Deligne: Théorie de Hodge I, Actes ICM Nice 1970, Gauthier-Villars 1971, I, p. 425-430.

[D 3] P. Deligne: Equations Différentielles à Points Singuliers Réguliers, Springer Lecture Notes 163 (1970).

[D 4] P. Deligne: Théorie de Hodge II, Publ. Math. I.H.E.S. 40 (1972), 5-57.

[D 5] P. Deligne: Théorie de Hodge III, Publ. Math. I.H.E.S. 44 (1974), 5-78.

[D 6] P. Deligne: Valeurs de fonctions L et périodes d'intégrales, Proc. Symp. Pure Math. 33, 2 (1979), 313-342.

[D 7] P. Deligne: Poids dans la cohomologie des variétés algébriques, Actes ICM Vancouver 1974, I, p. 79-85.

[D 8] P. Deligne: La conjecture de Weil I, Publ. Math. I.H.E.S. 43 (1974), 273-308.

[D 9] P. Deligne: La conjecture de Weil II, Publ. Math. I.H.E.S. 52 (1981), 313-428.

[D 10] P. Deligne: Letter to Soulé, January 20, 1985.

[D 11] P. Deligne: Le groupe fondamental de la droite projective moins trois points, Galois groups over $\mathbb{Q}$, MSRI Publications, Springer 1989, 79-297.

[DMOS] P. Deligne, J. Milne, A. Ogus, K. Shih: Hodge Cycles, Motives and Shimura Varieties, Springer Lecture Notes 900 (1982).

[DV] A. Douady, J.-L. Verdier, et al.: Séminaire de Géometrie Analytique, asterisque 36-37 (1976).

[EV] H. Esnault, E. Viehweg: Deligne-Beilinson cohomology; Beilinson's Conjectures on Special Values of L-Functions, Perspectives in Mathematics 4, Academic Press 1988, 43-91.

[Fa] G. Faltings: Crystalline cohomology and p-adic Galois representations, preprint 1988.

[FW] G. Faltings, G. Wüstholz, et al.: Rational Points, Sem. Bonn/Wuppertal 1983/84, Vieweg, Braunschweig/Wiesbaden 1984.

[Fo] J.-M. Fontaine: Sur certains types de représentations p-adiques du groupe de Galois d'un corps local; construction d'un anneau de Barsotti-Tate, Ann. of Math. 115 (1982), 529-577.

[FM] J.-M. Fontaine, W. Messing: p-adic periods and p-adic etale cohomology; Current Trends in Arithmetical Algebraic Geometry, Contemporary Mathematics 67, A.M.S., Providence 1987, 179-207.

[Fr] E. Friedlander: Etale K-theory II. Connections with algebraic K-theory, Ann. Sci. ENS (4) 15 (1982), 231-256.

[Fu] W. Fulton: Rational equivalence on singular varieties, Publ. Math. I.H.E.S. 45 (1975), 147-167.

[Gi] H. Gillet: Riemann-Roch theorems for higher algebraic K-theory, Advances in Math. 40 (1981), 203-289.

[Gr] A. Grothendieck: Hodge's general conjecture is false for trivial reasons, Topology 8 (1969), 299-303.

[Har 1] G. Harder: Die Kohomologie S-arithmetischer Gruppen über Funktionenkörpern, Invent. Math. 42 (1977), 135-175.

[Harts] R. Hartshorne: On the de Rham cohomology of algebraic varieties, Publ. Math. I.H.E.S. 45 (1976), 5-99.

[HL] M. Herrera, D. Lieberman: Duality and the de Rham cohomology of infinitesimal neighborhoods, Invent. Math. 13 (1971),97-124.

[Hir] H. Hironaka: Resolution of singularities of an algebraic variety over a field of characteristic zero, Ann. of Math. 79 (1964), 109-326.

[Ho 1] G. Hochschild: Cohomology of algebraic linear groups, Illinois J. Math. 5 (1961), 492-519.

[Ho 2] G. Hochschild: Lie algebra cohomology and affine algebraic groups, Illinois J. Math. 18 (1974), 170-176.

[J 1] U. Jannsen: Continuous étale cohomology, Math. Ann. 280 (1988), 207-245.

[J 2] U. Jannsen: Deligne homology, Hodge-D-conjecture, and motives; Beilinson's Conjectures on Special Values of L-Functions, Perspectives in Mathematics 4, Acadamic Press 1988, 305-372.

[J 3] U. Jannsen: On the ℓ-adic cohomology of varieties over number fields and its Galois cohomology; Galois groups over $\mathbb{Q}$, MSRI Publications, Springer 1989, 315-360.

[J 4] U. Jannsen: The derived category of Q_ℓ-sheaves, in preparation.

[J 5] U. Jannsen: On torsion in higher Chow groups, in preparation.

[Kl] S. Kleiman: Motives; Algebraic Geometry Oslo 1970, Wolters-Noordhoff, Groningen 1972, 53-82.

[La] S. Lang: Fundamentals of Diophantine Geometry, Springer Verlag, New York 1983.

[Li 1] S. Lichtenbaum: Values of zeta-functions at non-negative integers, Springer Lecture Notes 1068 (1984), p. 127-138.

[Li 2] S. Lichtenbaum: The construction of weight-two arithmetic cohomology, Invent. Math. 88 (1987), 183-215.

[MacL] S. Mac Lane: Homology, Springer Berlin-Göttingen-Heidelberg 1963.

[Ma] Y. Manin: Correspondences, motives, and monoidal transformations, Mat. Sb. 77 no. 4 (1968), 475-507 (russian).

[Mi] J. Milne: Etale Cohomology, Princeton Mathematical Series 33, Princeton Univ. Press 1980.

[MS 1] A. Merkurjev, A. Suslin: K-cohomology of Severi-Brauer varieties and norm residue morphism, Math. USSR Izv. 21 (1983), 307-340.

[MS 2] A. Merkurjev, A. Suslin: On the K_3 of a field, preprint Leningrad 1987.

[Mu] D. Mumford: Rational equivalence of 0-cycles on surfaces, J. Math. Kyoto Univ. 9 (1968), 195-204.

[N] M. Nagata: Imbedding of an abstract variety in a complete variety, J. Math. Kyoto Univ. 2 (1962), 1-10.

[P] H. Pohlmann: Algebraic cycles on abelian varieties of complex multiplication type, Ann. of Math. 88 (1968), 161-180.

[Q 1] D. Quillen: Higher algebraic K-theory I, Springer Lecture Notes 341 (1973), p. 85-147.

[Q 2] D. Quillen (notes by D. Grayson): Finite generation of K-groups of a curve over a finite field, Springer Lecture Notes 966 (1982), p. 69-90.

[RZ] M. Rapoport, Th. Zink: Über die lokale Zetafunktion von Shimuravarietäten. Monodromiefiltration und verschwindende Zyklen in ungleicher Charakteristik, Invent. Math. 68, (1982), 21-101.

[Ro 1] A. Roitman: Rational equivalences of zero-cycles, Math. USSR Sb. 18 (1974), 571-588.

[Ro 2] A. Roitman: The torsion in the group of zero cycles modulo rational equivalence, Ann. Math. 111 (1980), 553-569.

[SR] N. Saavedra Rivano: Catégories Tannakiennes, Springer Lecture Notes 265 (1972).

[Scha] N. Schappacher: Periods of Hecke Characters, Springer Lecture Notes 1301 (1988).

[Sche] V. Schechtman: Algebraic K-theory and characteristic classes, Trudy MMO (Moscow Math. Soc.) 45 (1983), 237-264 (russian).

[Schoe] C. Schoen: Letter to the author, April 30, 1987.

[Sch 1] A. Scholl: Motives for modular forms, preprint Durham 1987.

[Sch 2] A. Scholl: Remarks on special values of L-functions, preprint 1988.

[Se 1] J.-P. Serre: Abelian ℓ-adic representations and elliptic curves, Benjamin, New York 1968.

[Se 2] J.-P. Serre: Représentations ℓ-adiques, Kyoto Int. Symp. on Algebraic Number Theory, Japan Soc. for the Promotion of Science 1977, 177-193.

[Sou 1] C. Soulé: K-théorie des anneaux d'entiers de corps de nombres et cohomologie étale, Invent. Math. 55 (1979), 251-295.

[Sou 2] C. Soulé: On higher p-adic regulators, Springer Lecture Notes 854 (1981), p. 372-401.

[Sou 3] C. Soulé: Opérations en K-théorie algébrique, Canad. J. Math. 37 (1985), 488-550.

[Sou 4] C. Soulé: Groupes de Chow et K-théorie de varietés sur un corps fini, Math. Ann. 268 (1984), 317-345.

[Su 1] A. Suslin: Stability in algebraic K-theory, Springer Lecture Notes 966 (1982), p. 304-333.

[Su 2] A. Suslin: Homology of GL_n, characteristic classes and Milnor K-theory, Springer Lecture Notes 1046 (1984), p. 357-375.

[Su 3] A. Suslin: Torsion in K_2 of fields, preprint Leningrad 1982 (compare MS 1).

[T 1] J. Tate: Algebraic cycles and poles of zeta functions; Arithmetical Algebraic Geometry, ed. O.F.G. Schilling, Harper and Row, New York 1965, p. 93-100.

[T 2] J. Tate: Endomorphisms of abelian varieties over finite fields, Invent. Math. 2 (1966), 134-144.

[Th] R. Thomason: Absolute cohomological purity, Bull. Soc. Math. France 112 (1984), 397-406.

[D.E.] J. Giraud et al.: Dix exposés sur la cohomologie des schémas, North-Holland Publish. Co., Amsterdam 1968.

[SGA 4] M. Artin, A. Grothendieck, J.L. Verdier, et al.: Théorie des Topos et Cohomologie Etale des Schémas, Springer Lecture Notes 269, 270, 305 (1972/73).

[SGA $4\frac{1}{2}$] P. Deligne et al.: Cohomologie Etale, Springer Lecture Notes 569 (1977).

[SGA 5] A. Grothendieck et al.: Cohomologie ℓ-adique et Fonctions L, Springer Lecture Notes 589 (1977).

[EGAIV] A. Grothendieck, J. Dieudonné: Eléments de Géométrie Algébrique IV, Etude Locale des schémas et de morphismes de schémas, Publ. Math. I.H.E.S. 28 (1966).

[SGA 6] P. Berthelot, A. Grothendieck, L. Illusie et al.: Théorie des Intersections et Théorème de Riemann-Roch, Springer Lecture Notes 225 (1971).

Notations

$\mathbb{Z}$ and $\mathbb{Z}_\ell$ are the rings of rational and ℓ-adic integers, respectively.

$\mathbb{Q}$, $\mathbb{R}$, $\mathbb{C}$, $\mathbb{Q}_\ell$ are the fields of rational, real, complex, and ℓ-adic numbers, respectively.

$\mathbb{F}_q$ is the field with q elements.

$R^\times$ is the multiplicative group of a ring R.

A^G, for a group G and a G-module A, is the fixed module: $A^G = \{a \in A \mid ga = a$ for all $g \in G\}$.

If X is a variety over a field k, then $X(k)$ is the set of k-rational points, $X \times_k k'$ (or $X \times_{k,\sigma} k'$) is the base extension via a field extension $\sigma : k \hookrightarrow k'$, and $\mathrm{Pic}(X)$ is the Picard group of X. If X is smooth, $\underline{\mathrm{Pic}}^0(X)$ is the Picard variety, and $\mathrm{Pic}^0(X) = \underline{\mathrm{Pic}}^0(X)(k)$.

$H^i(X,A) = H^i(X(\mathbb{C}),A)$, for an abelian group A and a variety X over $\mathbb{C}$, is the singular (Betti) cohomology of the topological space $X(\mathbb{C})$ with coefficients in A. If F is a sheaf for the analytic topology on X (i.e., a sheaf on the associated complex analytic space X^{an}), then $H^i(X,F) = H^i(X^{an},F)$ is the analytic sheaf cohomology. Since canonically $H^i(X^{an},\underline{A}) = H^i(X(\mathbb{C}),A)$ for the constant sheaf $\underline{A}$ associated to A, we mostly write A again for $\underline{A}$ without risk of confusion. $H^i_B(X,\mathbb{Z})(j) = H^i(X,\mathbb{Z}(j))$ is the j-fold Tate twist of the integral (mixed) Hodge structure given by $H^i(X,\mathbb{Z})$.

$H^i_{\acute{e}t}(X,F)$ denotes étale cohomology of an étale sheaf F on a scheme X (see, e.g., [Mi]). If ℓ is a prime invertible on X, we let as usual $H^i_{\acute{e}t}(X,\mathbb{Z}_\ell(j)) = \varprojlim_n H^i_{\acute{e}t}(X,\mu_{\ell^n}^{\otimes j})$, where μ_{ℓ^n} is the sheaf of ℓ^n-th roots of unity, and put $H^i_{\acute{e}t}(X,\mathbb{Q}_\ell(j)) = H^i_{\acute{e}t}(X,\mathbb{Z}_\ell(j)) \otimes_{\mathbb{Z}_\ell} \mathbb{Q}_\ell$. For these groups we often omit the index "ét". Similar notions apply for more general $\mathbb{Z}_\ell$- or $\mathbb{Q}_\ell$ - "sheaves", for which we refer to loc. cit. or [SGA 5].

$\pi_1(U,\bar{\eta})$, for a scheme U and a geometric point $\bar{\eta} \longrightarrow U$, is the algebraic fundamental group (cf. [Mi] I § 5).

Cone (α) is the cone of a morphism $\alpha : A^{\cdot} \longrightarrow B^{\cdot}$ of complexes, $A^{\cdot}[r]$ is the r–fold shift of $A^{\cdot}$ (cf. [Mi] p. 167, 174).

As usual, $D^b(A)$ is the derived category of bounded complexes in an abelian category A. We also use other standard notations connected with derived categories (like Rf_*, $Rf^!$, $\otimes^L$, RHom, ...), which can be found, e.g., in [SGA 4] XVII and XVIII.

tr. deg(k) is the degree of transcendence of a field k over its prime field.

Other notations are introduced at the following places:

$H^n_{DR}(X)$, $H^n_{DR}(U)$	1,25	$\overset{\circ}{\underline{V}}_k$	25
$H^n_\ell(X)$, $H^n_\ell(U)$	1,32	Ω^p_X, $\Omega^p_X\langle Y\rangle$	25,26
$H^n_\sigma(X)$, $H^n_\sigma(U)$	1,32	$Y^{(q)}$ (two meanings)	27,76
$I_{\infty,\sigma}$	1,10,33	σU^{an}	32
$I_{\ell,\bar{\sigma}}$	1,10,34	$H^n(U)$	35
σX, σU	2,32	$\bar{U}, \bar{X}$	36,57
$\underline{MR}_k$	9	$\underline{MM}_k$, $\underline{M}_k$	43,46
H_{DR}, H_ℓ, H_σ	10	$G(\sigma)$, $MG(\sigma)$, $\underline{Rep}\ G$	49
H	10,43,46	$U(\sigma)$, sp_ℓ, Msp_ℓ	50
W_m	10,83,87	$CH^r(X)$	57
Gr^W_m, $\underline{R}_k$	12	$c\ell^r_\ell$, Γ_ℓ, Γ_{DR}, $\Gamma_{\mathcal{H}}$	57,62
$H \otimes H'$, $\underline{Hom}(H,H')$	12,13	$c\ell^r_{DR}$, $c\ell^r_\sigma$, $c\ell^r_{\mathcal{AH}}$, $\Gamma_{\mathcal{AH}}$	58,59,62
$\underline{1}$ (identity object)	13,80	$K_m(X)$, $ch_{i,j}$	65
Γ	14,81,125	$K_m(U)^{(j)}$, $H^i_{\mathcal{M}}(U,\mathbb{Q}(j))$	67
$H^\vee$	15	$H^i_{\mathcal{D}}$	68
$\underline{Vec}_F$	15,125	H^i_{cont}	70
$H \times_k k'$, $R_{k'/k}$	16,75	$\hat{A} = \varprojlim_n A/\ell^n$	70,71
$Ind_{G_{k'}}^{G_k}$	17	N^iH^n	76,162
$\underline{1}(n)$, $H(n)$	17	$\kappa(x)$	76

Index

LECTURE NOTES IN MATHEMATICS

Edited by A. Dold and B. Eckmann

Some general remarks on the publication of monographs and seminars

In what follows all references to monographs, are applicable also to multiauthorship volumes such as seminar notes.

§1. Lecture Notes aim to report new developments - quickly, informally, and at a high level. Monograph manuscripts should be reasonably self-contained and rounded off. Thus they may, and often will, present not only results of the author but also related work by other people. Furthermore, the manuscripts should provide sufficient motivation, examples and applications. This clearly distinguishes Lecture Notes manuscripts from journal articles which normally are very concise. Articles intended for a journal but too long to be accepted by most journals, usually do not have this "lecture notes" character. For similar reasons it is unusual for Ph.D. theses to be accepted for the Lecture Notes series.

Experience has shown that English language manuscripts achieve a much wider distribution.

§2. Manuscripts or plans for Lecture Notes volumes should be submitted either to one of the series editors or to Springer-Verlag, Heidelberg. These proposals are then refereed. A final decision concerning publication can only be made on the basis of the complete manuscripts, but a preliminary decision can usually be based on partial information: a fairly detailed outline describing the planned contents of each chapter, and an indication of the estimated length, a bibliography, and one or two sample chapters - or a first draft of the manuscript. The editors will try to make the preliminary decision as definite as they can on the basis of the available information.

§3. Lecture Notes are printed by photo-offset from typed copy delivered in camera-ready form by the authors. Springer-Verlag provides technical instructions for the preparation of manuscripts, and will also, on request, supply special staionery on which the prescribed typing area is outlined. Careful preparation of the manuscripts will help keep production time short and ensure satisfactory appearance of the finished book. Running titles are not required; if however they are considered necessary, they should be uniform in appearance. We generally advise authors not to start having their final manuscripts specially tpyed beforehand. For professionally typed manuscripts, prepared on the special stationery according to our instructions, Springer-Verlag will, if necessary, contribute towards the typing costs at a fixed rate.

The actual production of a Lecture Notes volume takes 6-8 weeks.

.../...

§4. Final manuscripts should contain at least 100 pages of mathematical text and should include

- a table of contents
- an informative introduction, perhaps with some historical remarks. It should be accessible to a reader not particularly familiar with the topic treated.
- a subject index; this is almost always genuinely helpful for the reader.

§5. Authors receive a total of 50 free copies of their volume, but no royalties. They are entitled to purchase further copies of their book for their personal use at a discount of 33.3 %, other Springer mathematics books at a discount of 20 % directly from Springer-Verlag.

Commitment to publish is made by letter of intent rather than by signing a formal contract. Springer-Verlag secures the copyright for each volume.